Strategies for Waste Disposal and Pollution Control

Strategies for Waste Disposal and Pollution Control

Edited by **Sheryl McMillan**

SYRAWOOD
PUBLISHING HOUSE

New York

Published by Syrawood Publishing House,
750 Third Avenue, 9th Floor,
New York, NY 10017, USA
www.syrawoodpublishinghouse.com

Strategies for Waste Disposal and Pollution Control
Edited by Sheryl McMillan

International Standard Book Number: 978-1-68286-012-0 (Hardback)

Printed in the United States of America.

Contents

Preface

Over the recent decade, advancements and applications have progressed exponentially. This has led to the increased interest in this field and projects are being conducted to enhance knowledge. The main objective of this book is to present some of the critical challenges and provide insights into possible solutions. This book will answer the varied questions that arise in the field and also provide an increased scope for furthering studies.

Scientists and research organizations across the globe are trying to build novel strategies for waste disposal and treatment of industrial effluents. The ever growing need to control pollution and dispose waste in an eco-friendly manner is the reason that has fuelled the research in this field in recent times. The chapters included in this book contain some of the vital pieces of work being conducted across the world, on various topics related to waste management and pollution control like evaluation and management of environmental risk and safety, waste treatment and recycling, waste water treatment, etc. It is bound to provide innovative insights to the students and researchers engaged in this field.

I hope that this book, with its visionary approach, will be a valuable addition and will promote interest among readers. Each of the authors has provided their extraordinary competence in their specific fields by providing different perspectives as they come from diverse nations and regions. I thank them for their contributions.

Editor

A Simple and Objective Carbon Footprint Analysis for Alternative Transportation Technologies

R. A. Dunlap[1]

[1] Department of Physics and Atmospheric Science, Dalhousie University, Halifax, Nova Scotia, Canada

Correspondence: R. A. Dunlap, Department of Physics and Atmospheric Science, Dalhousie University, Halifax, Nova Scotia B3H 4R2, Canada. E-mail: dunlap@fizz.phys.dal.ca

Abstract

A simple and straightforward method of analyzing the net carbon dioxide emissions that result from different alternative transportation technologies is presented. Results are shown for three different non-fossil fuel transportation technologies; battery electric vehicles, hydrogen fuel cell vehicles and hydrogen internal combustion vehicles. These results are compared with the carbon emissions of a traditional gasoline powered internal combustion engine vehicle. Battery electric vehicles are shown to have a significantly lower carbon footprint than gasoline vehicles. Fuel cell vehicles are comparable to gasoline vehicles while hydrogen internal combustion vehicles produce substantially more carbon dioxide. The importance of electricity generating infrastructure is discussed.

Keywords: carbon dioxide emissions, greenhouse gases, battery electric vehicles, fuel cell vehicles, hydrogen vehicles

1. Introduction

The development of non-fossil fuel based energy technologies is one of most important and difficult tasks facing humanity. Energy sources for transportation applications are a particularly challenging aspect of establishing a carbon-free energy economy for the future as they must be portable and robust. A consideration of the relevant criteria for some alternative transportation energy technologies has been presented in (e.g., Kraushaar & Ristinen, 1993; Kruger, 2006; Vanek & Albright, 2008). Suitable transportation technologies for widespread implementation must provide environmental advantages over traditional fossil fuel technologies, specifically with regard to total CO_2 emissions, that is, their carbon footprint. A detailed and complete CO_2 emission analysis is generally complex (see e.g. Korchinski, 2007) and must include a lifecycle analysis of all materials and involved. For this reason, a direct comparison of the different available transportation energy technologies is difficult. As well, economic considerations are important in determining viability of various technologies. Net efficiency is a critical factor in determining cost and while certain technologies may have minimal environmental impact, the efficiency is unacceptably low (e.g. hydrogen internal combustion vehicles, as discussed below). Some technologies (such as battery electric vehicles) have become more cost effective in recent years but widespread use will require the implementation of a suitable infrastructure and public acceptance of possible drawbacks (e.g. limited range and/or long recharge times).

The current paper deals with the implementation of a simple approach for understanding CO_2 emissions. While the details of lifecycle analysis are not considered, the method does provide a quantitative technique for comparing the environmental impact of different transportation energy technologies. Battery electric vehicles (BEV's), fuel cell vehicles and hydrogen internal combustion engine (H_2 ICE) vehicles are compared with a traditional gasoline internal combustion engine vehicle in terms of expected CO_2 emissions per kilometer. This analysis provides an appreciation for the factors that are relevant in assessing the environmental impact of different technologies and emphasizes that the common perception of environmentally friendly technologies can be misleading. It also provides a means for assessing the importance of electricity generating infrastructure in evaluating the carbon footprint of transportation methods. The significance of energy technologies that are prevalent in different countries is discussed.

2. Analysis of CO_2 Emissions

The goal of the present analysis is to calculate the CO_2 emissions per kilometer for different transportation technologies in an objective manner that allows for direct comparison of these technologies and provides the student with an understanding of the relevant factors in such an analysis. The mass of CO_2 emitted per kilometer traveled may be expressed as

$$\frac{kg(CO_2)}{km} = \frac{kg(CO_2)}{E_p} \times \frac{E_p}{E_w} \times \frac{E_w}{km} \qquad (1)$$

where E is energy. The subscript "p" refers to primary energy and the subscript "w" refers to energy delivered to the vehicle's wheels. The first term on the right hand side of the equation represents the amount of CO_2 generated per unit of primary energy consumed (expressed in the present paper in MJ). The second term is the inverse of the conversion efficiency from primary energy to wheel energy. The final term on the right hand side is the average energy to the wheels needed to move the vehicle 1 km. It is easy to see that the product of the terms on the right hand side of the equation will reduce to the expression on the left hand side. However, it is convenient to write the right hand side of the equation in this way as it illustrates the importance the various factors involved and provides a means of readily undertaking a quantitative analysis of CO_2 emissions. In the case where different primary energy sources are used to ultimately provide energy for the same transportation technology, then Equation (1) can be written as a sum over the relevant sources as

$$\frac{kg(CO_2)}{km} = \sum_i f_i \left[\frac{kg(CO_2)}{E_p} \right]_i \left[\frac{E_p}{E_w} \right]_i \frac{E_w}{km} \qquad (2)$$

where f_i is the relative fraction of energy from a particular primary source. For example, equation (2) may be applied to a battery electric vehicle where the electricity use to charge the batteries comes from a variety of sources, e.g. coal, natural gas, nuclear, hydroelectric, etc. In this case there would be a term "i" in the sum for each of these primary sources. The conversion of primary energy to electricity is an important factor in determining the overall effectiveness of an energy infrastructure and has been the subject of a number of studies. Detailed analyses are important for a consideration of specific situations such as those presented for California by McCollum et al. (2012) and Poland by Budzianowski (2011).

The analysis presented in the current work does not consider the carbon footprint associated with vehicle manufacture and disposal or the infrastructure associated with fuel transportation or marketing, although these effects will tend to average out somewhat among the different technologies. The present analysis is, therefore, a reasonable comparison of different vehicle technologies and because of its simple and quantitative nature, it represents a useful pedagogical approach and gives students an understanding of the importance of an objective scientific analysis of the environmental aspects of non-fossil fuel energy. The analysis of the various terms on the right hand side of Equation (1) is considered below.

2.1 Analysis of kg(CO₂)/(E_p)

For a gasoline internal combustion engine (ICE) powered vehicle, gasoline is very close to a primary energy source, while for battery electric vehicles (BEV's) or hydrogen powered vehicles, the primary energy sources are first used to produce electricity, which is then stored for vehicle use. Thus $kg(CO_2)/(E_p)$ depends on the way in which electricity is produced, that is which primary energy sources are used. As an example, a rough breakdown of present electricity production in the U.S. is shown in Table 1.

Table 1. Breakdown of electricity production in the United States in 2011 as reported on the U.S. Energy Information Administration website (EIA, 2012)

fuel	% U.S. electricity[a]
coal	42.3
natural gas [b]	25.1
petroleum [c]	0.7
non-fossil fuel	31.9

Notes: (a) does not include about 0.2% miscellaneous sources, (b) includes other fossil-fuel derived gases, (c) includes petroleum derived products.

The CO_2 emission per unit energy (MJ) for different fossil fuels is obtained from an analysis of their energy content and is shown in Table 2. Budzianowski (2012) gives values of carbon emissions from primary energy sources as 24.5 g(C)/MJ [grams of carbon per MJ] for coal/peat, 14.7 g(C)/MJ for natural gas and 17.5 g(C)/MJ for oil. These are equivalent (in units used in the present work) to 0.090 kg(CO_2)/MJ, 0.054 kg(CO_2)/MJ and 0.064 kg(CO_2)/MJ, respectively. Given the variability in the chemical content of coal and heavy hydrocarbons, these literature values are consistent with the values presented in Table 2. The possible effect of carbon sequestration from fossil fuel generation is not considered and the CO_2 emissions from non-fossil fuel sources (i.e. nuclear, hydroelectric and alternative energy sources) is considered to be zero, i.e. the term [kg(CO_2)/E_p]$_i$ = 0 for nuclear, hydroelectric, etc. These assumptions will tend to cancel each other somewhat. While ignoring these factors will tend to present an overly optimistic view of non-fossil fuel energy, these effects are relatively small (at present), difficult to quantify and do not alter the basic conclusions of pedagogical value of the present approach.

Table 2. The CO_2 emission per MJ for different fossil fuels

fossil fuel	kg(CO_2)/E_p [kg/MJ]
carbon (~coal)	0.11
natural gas (methane, CH_4)	0.055
heavy hydrocarbons (>6 C/molecule)	0.069

As E_p/E_w is expected to be similar for all fossil fuel generated electricity (i.e. it is limited by the thermodynamic efficiency of a heat engine) it is suitable to use an average value of the CO_2 emissions from these generating methods in Equation (2) and a corresponding total value of f_i for fossil fuel generation. The average kg(CO_2)/E_p is obtained from the values in Table 2 weighted by the percentages in Table 1. For fossil fuel generation of electricity in the U.S. as discussed above this analysis will give

$$
\begin{aligned}
<\text{kg}(CO_2)/E_p> &= 0.11 \text{ kg}(CO_2)/\text{MJ} \times (0.423) \\
&+ 0.055 \text{ kg}(CO_2)/\text{MJ} \times (0.251) \\
&+ 0.069 \text{ kg}(CO_2)/\text{MJ} \times (0.007) \\
&= 0.061 \text{ kg}(CO_2)/\text{MJ}
\end{aligned}
\tag{3}
$$

This value is the appropriate average (for the United States) for all alternative energy vehicles that utilize electricity and a portable electricity storage mechanism (e.g. batteries or hydrogen). This calculation will, of course, be different for different locations worldwide as the distribution of primary energy sources used to produce electricity varies. National comparisons for some countries are discussed below.

2.2 Analysis of $(E_p)/(E_w)$

The quantity E_p/E_w is a measure of the amount of primary energy needed to provide one unit of useable energy to the vehicle's wheels. It is the inverse of the overall efficiency (as a fraction) of the energy conversion processes involved and is given as

$$
\frac{E_p}{E_w} = \frac{100}{\text{efficiency(\%)}}
\tag{4}
$$

There is often considerable uncertainty in estimating the efficiency of various energy conversion processes and there are, as well, indirect energy "costs", such as (e.g.) energy required for infrastructure development and fuel transportation. The estimates for the efficiency of the different energy conversion processes involved in converting primary energy to energy at the wheels for gasoline internal combustion vehicles and some alternative technologies are discussed below. The estimated efficiencies presented below are approximate values for the processes involved. More detailed analysis can include specific information about particular cases, e.g. specific vehicles, generating facilities, etc. if appropriate.

2.2.1 Gasoline Internal Combustion Engine

Table 3 gives the efficiency of the energy conversion process relevant to a gasoline internal combustion engine (ICE) vehicle. The efficiency of converting energy stored in gasoline to mechanical energy at the vehicle's wheels is limited by the thermodynamic efficiency of a heat engine. The value expressed in Table 3 is a typical net efficiency for an automobile engine. It is assumed that conversion of primary energy (i.e. crude oil) to gasoline has a high efficiency and gasoline can, therefore, be treated as a primary energy source.

Table 3. Efficiency analysis for gasoline powered internal combustion engine vehicle showing net efficiency for conversion of primary energy (gasoline) to mechanical energy delivered to the vehicle's wheels

process	efficiency
fossil fuel → mechanical	17%
net efficiency	**17%**

2.2.2 Battery Electric Vehicles

The relevant efficiencies for a battery electric vehicle (BEV) using a fossil fuel primary energy source are shown in Table 4. The efficiency of primary energy to electricity is taken to be the average efficiency of a thermal generating station burning (e.g.) coal. Estimated efficiency for conversion of electrical energy to mechanical energy at the vehicle's wheels accounts for battery storage efficiency and electric motor efficiencies.

Table 4. Efficiency analysis for battery electric vehicle showing net efficiency for conversion of primary energy (fossil fuel) to mechanical energy delivered to the vehicle's wheels

process	efficiency
fossil fuel → electricity	40%
electricity → mechanical	85%
net efficiency	**34%**

2.2.3 Hydrogen Internal Combustion Engine

Table 5 provides information about the typical route of hydrogen production and utilization for a hydrogen ICE. Hydrogen gas, at STP, is produced by electrolysis of water and is then compressed or liquefied for vehicle use. Internal combustion engine efficiency follows along the lines of the efficiency of a gasoline engine and is limited by thermodynamic factors.

Table 5. Efficiency analysis for hydrogen powered internal combustion engine vehicle showing net efficiency for conversion of primary energy (fossil fuel) to mechanical energy delivered to the vehicle's wheels. CHG = compressed hydrogen gas, LH_2 = liquid hydrogen

process	efficiency
fossil fuel → electricity	40%
electricity → hydrogen gas	70%
hydrogen gas → CHG/LH_2	80%
CHG/LH_2 → mechanical	17%
net efficiency	**4%**

2.2.4 Fuel Cell Vehicles

The production and storage of hydrogen is described above. The efficiency of converting hydrogen to electricity (i.e. the fuel cell efficiency) is the efficiency of state-of-the art polymer electrolyte membrane (PEM) fuel cells which are typically most appropriate for vehicle use. Fuel cell vehicle efficiency is summarized in Table 6.

Table 6. Efficiency analysis for hydrogen fuel cell powered vehicle showing net efficiency for conversion of primary energy (fossil fuel) to mechanical energy delivered to the vehicle's wheels

process	efficiency
fossil fuel → electricity	40%
electricity → hydrogen gas	70
hydrogen gas → CHG	80
CHG → electricity	70
electricity → mechanical	90
net efficiency	**14%**

2.3 Analysis of (MJ)$_w$/km

The energy per unit distance required for a vehicle can be determined on the basis of the characteristics of current gasoline powered vehicles. Specifically, the energy content of the fuel consumed per unit distance and the efficiency of the process are needed. In the case of the gasoline powered vehicle it is important to use the efficiency as presented in Table 3. In this case the gasoline powered vehicle efficiency will cancel out in Equation (2). Table 7 gives fuel consumption figures for a range of automobiles from major manufacturers. The values for mpg (miles per U.S. gallon) are combined city/highway figures as reported by the United States Environmental Protection Agency (2012). The fuel consumption in litres per km can be determined from the published values in mpg according to the conversion

$$\frac{litres}{km} = \frac{2.35 \ (litres \times mpg/km)}{mpg} \tag{5}$$

The energy requirement in MJ delivered to the wheels per km can be determined from the energy content of gasoline 34.8 MJ/litre and the propulsion efficiency as

$$\frac{E_w}{km} = \frac{litres}{km} \times \frac{MJ}{litre} \times efficiency \tag{6}$$

Calculated values for the various vehicles in Table 7 are given in the last column. These cover a range of values that is approximately a factor of 2 with an average around 0.55 MJ/km for a typical family sedan. This typical value will be used for further analysis, and Table 7 gives the range of values expected for different vehicles.

Table 7. U.S. Environmental Protection Agency (EPA) combined city/highway mileage ratings for selected 2012 automobiles with standard engine option and automatic transmission. The energy content of gasoline is 34.8 MJ per litre and the average efficiency of an internal combustion engine was assumed to be 17%. Values are calculated from data on the EPA website (EPA, 2012)

Vehicle (make/model)	mpg (US)	litres/km	E_w/km [MJ/km]
smart fortwo	36	0.065	0.39
Chevrolet Sonic	33	0.071	0.42
Hyundai Sonata	28	0.084	0.50
Toyota Camry (V6)	25	0.094	0.56
BMW 535xi GT	21	0.112	0.66
Mercedes Benz S550	18	0.131	0.77

3. National Comparisons of kg(CO_2)/km

The data as presented in the previous section can now be substituted into Equation (2) to calculate the CO_2 emissions for the various transportation technologies discussed here. For the electricity generating technologies in current use in the United States the calculated values of kg(CO_2)/km are given in Table 8.

Table 8. Calculated values for the CO_2 emission per km for different transportation technologies in the United States

Technology	$<kg(CO_2)/E_p>$ [kg/MJ]	E_p/E_w	E_w/km [MJ/km]	$kg(CO_2)$/km
gasoline ICE	0.069	5.9	0.55	0.22
BEV	0.061	2.9	0.55	0.097
H_2 ICE	0.061	25	0.55	0.83
H_2 fuel cell	0.061	7.1	0.55	0.24

This analysis shows that the net environmental impact of battery electric vehicles in the United States is quite positive (compared with gasoline vehicles). Fuel cell vehicles are about neutral, while H_2 internal combustion engine vehicles have a very negative environmental effect. Improvements to the environmental impact of alternative fuel vehicles can be made by increasing energy conversion efficiencies (i.e. decreasing E_p/E_w and increasing vehicle efficiencies (i.e. decreasing E_w/km). However, it is through the shift in electricity generating technology away from fossil fuels (i.e. the reduction of $<kg(CO_2)/E_p>$), that the most substantial improvements can be made.

The effects of electric generating technology that is prevalent in certain countries are illustrated in Table 9. The factor $<kg(CO_2)/E_p>$ is a multiplicative factor in Equation (2) for alternative fuel vehicles and the resulting $kg(CO_2)$/km values scale correspondingly. The value for the world average as calculated in the present work (0.060 $kg(CO_2)$/MJ) is consistent with the analysis of world primary energy consumption as presented by the International Energy Agency (IEA, 2011) which gives an average of 15.4 g(C)/MJ (or 0.056 $kg(CO_2)$/MJ). While the world average is essentially identical to the value for the United States, some countries have energy policies that make electricity production more environmentally advantageous. This is directly a result of the lower than (worldwide) average of fossil fuels used in electricity generation. Two examples are clearly identifiable in Table 9, Canada and France, where electricity is produced primarily by CO_2-free methods; hydroelectricity and nuclear energy, respectively. In Canada there is a reduction of about a factor of 3 in $kg(CO_2)$/km production for alternative fuel vehicles compared with the U.S. for the same type of alternative technology vehicle. In France there is about a factor of 8 reduction. This means that in Canada BEV's and fuel cell vehicles are clearly advantageous over gasoline vehicles and in France even H_2 ICE vehicles come out ahead of gasoline ICE vehicles in net CO_2 emissions. Thus it is clear that, at least from a greenhouse gas emissions standpoint, alternative transportation technologies are more environmentally attractive in certain countries.

Table 9. National comparisons for some countries and world average for electricity production for 2008 from the International Energy Agency website (IEA, 2012). Calculated values of the average $<kg(CO_2)/E_p>$ as described above for alternative fuel vehicles are shown

nation	coal	natural gas	petroleum	non-fossil fuels	principal non-fossil fuel type	$<kg(CO_2)/E_p>$ [kg/MJ]
Australia	77.0	15.2	1.1	6.7	hydroelectric	0.094
Canada	17.2	6.4	1.5	74.9	hydroelectric	0.023
China	79.1	0.9	0.7	13.0	hydroelectric	0.088
France	4.7	3.8	1.0	90.5	nuclear	0.008
Germany	45.7	13.8	1.4	39.1	nuclear	0.059
India	68.6	9.9	4.1	17.4	hydroelectric	0.084
Japan	26.6	26.2	12.8	34.4	nuclear	0.052
U.K	32.6	45.5	1.6	20.3	nuclear	0.062
World average	41.0	21.0	5.5	32.5	hydroelectric	0.060

4. Conclusions

The environmental aspects of energy use are an important consideration and the viability of pursuing alternative technologies which have more serious environmental consequences than traditional fossil fuel technology is questionable, at least in the short term. The simple analysis presented in the current work shows that in the United States that the average CO_2 emissions for BEV's, H_2 ICE vehicles and H_2 fuel cell vehicles are 0.097 kg (CO_2)/km, 0.83 kg(CO_2)/km and 0.24 kg(CO_2)/km, respectively, compared with 0.22 kg(CO_2)/km for a conventional gasoline ICE vehicle. The average situation world wide is shown to be essentially the same as in the United States and shows that overall BEV's provide a clear environmental advantage, fuel cell vehicles are about neutral and H_2 ICE vehicles are environmentally counterproductive. It is clear that the current advantages of alternative energy for transportation purposes is directly related to electric generating methods and improvements in the carbon foot print of this infrastructure in most parts of the world would improve the attractiveness of alternative transportation technologies. The desirability of different energy technologies must also be considered on the basis of economic factors and resource availability factors but a positive environmental impact is a necessary condition for viability. A quantitative analysis of CO_2 emissions is typically not a component of the literature that most likely forms public opinion (see e.g. discussion of hydrogen vehicles by Transport Canada, 2012).

The present paper presents an objective method of analyzing the environmental impact (in terms of CO_2 emissions) for BEV's, fuel cell vehicles and H_2 ICE vehicles in comparison with traditional gasoline powered vehicles. This analysis provides a quantitative approach to understanding this problem that is based on sound physical principles and emphasizes the need for such a quantitative evaluation of the carbon footprint of alternative energy technologies. Given the popularity of sustainable energy related courses in university science and engineering faculties, such a quantitative approach can supplement discussions of alternative transportation technologies and provide students with an appreciation of the complexity of a thorough objective analysis.

References

Budzianowski, W. M. (2011). Can 'negative net CO_2 emissions' from decarbonised biogas-to-electricity contribute to solving Poland's carbon capture and sequestration dilemmas? *Energy, 36*, 6318-6325. http://dx.doi.org.ezproxy.library.dal.ca/10.1016/j.energy.2011.09.047

Budzianowski, W. M. (2012). Target for national carbon intensity of energy by 2050: A case study of Poland's energy system. *Energy, 46*, 575-581. http://dx.doi.org.ezproxy.library.dal.ca/10.1016/j.energy.2012.07.051

EIA. (2012). *U.S. Energy Information Administration*. Retrieved September 26, 2012, from http://www.eia.gov/electricity/monthly/epm_table_grapher.cfm?t=epmt_1_1

EPA. (2012). *Environmental Protection Agency*. Retrieved September 26, 2012, from http://www.fueleconomy.gov/feg/Find.do?action=sbsSelect

IEA. (2011). *Key World Energy Statistics*. Retrieved from http://www.iea.org/publications/freepublications/publication/key_world_energy_stats-1.pdf

IEA. (2012). *International Energy Agency*. Retrieved September 26, 2012, from http://www.iea.org/stats/index.asp

Korchinski, W. J. (2007). *Are hydrogen cars good for America*? Reason Foundation, Policy Study 363. Retrieved from http://reason.org/files/87189798b5d1d8ce1be8fb811a21a3d0.pdf

Kraushaar, J. A., & Ristinen, R. A. (1993). *Energy and Problems of a Technical Society* (2nd ed.). New York: Wiley.

Kruger, P. (2006). *Alternative Energy Resources: The Quest for Sustainable Energy*. Hoboken: Wiley.

McCollum, D., Yang, C., Yeh, S., & Ogden, J. (2012). Deep greenhouse gas reductions scenarios for California - strategic implications from the CA-TIMES energy economic systems model. *Energy Strategy Reviews, 1*, 19-32. http://dx.doi.org/10.1016/j.esr.2011.12.003

Transport Canada. (2012). Retrieved October 19, 2012, from http://www.tc.gc.ca/eng/programs/environment-etv-videos-hydrogen-eng-1954.htm

Vanek, F. M., & Albright, L. D. (2008). *Energy Systems Engineering: Evaluation and Implementation*. New York: McGraw Hill.

2

In-situ Decomposition of Trichloroethylene Using Electrochemical Treatment Method

Takuya Ito[1], Kazuyuki Yamada[1], Sigeru Kato[1], Hideki Suganuma[1], Akihiro Yamasaki[1], Seiichi Suzuki[1] &
Toshinori Kojima[1]

[1] Department of Materials and Life Science, Seikei University, Tokyo, Japan

Correspondence: Takuya Ito, Department of Materials and Life Science, Seikei University, 3-3-1
Kichijoji-kitamachi, Musashino-shi, Tokyo, Japan. E-mail: takuya.ito@st.seikei.ac.jp

Abstract

Trichloroethylene (TCE) has an excellent degreasing capacity, so it is often used as a solvent for dry cleaning, and is still used for removing grease from metallic parts and so on. However, its inappropriate handling caused contamination of soil. Recently, its toxicity and carcinogenicity to humans have been concerned. By these reasons, it is highly required to remediate the contaminated soils. In the present study, the possibility of application of electrochemical treatment method to the *in-situ* decomposition of TCE is examined because *in-situ* remediation is expected to be simple and inexpensive. The experiment in the aqueous systems was conducted as a basic examination. As a result of comparing experimental values under various stirring speeds with the theoretical value calculated from mass transfer coefficient, it turned out that TCE transferring from bulk to the electrode surface is accelerated by the radicals in the boundary film near the electrode surface. Hence the TCE decomposition rate is affected by the radical formation rate or radical concentration in the boundary film. In the experiment with the soils, the TCE decomposition rate was much smaller than that in the aqueous systems. Moreover, the influence of the voltage was not observed. Therefore, it turned out that the movement of TCE in the aqueous phase near the electrode surface was the rate-controlling step in the soils. Under the condition, the TCE decomposition rate was not affected by the particle size. Consequently, it turned out TCE is not transported by bulk flow but is mostly transfered by molecular diffusion in the soil.

Keywords: trichloroethylene, electrochemical treatment, *in-situ* decomposition, contamination of soil, electroosmosis

1. Introduction

Trichloroethylene (TCE) has an excellent decreasing capacity, so it is often used as a solvent for dry cleaning, and is still used for removing grease from metallic parts and so on. However, its inappropriate handling caused contamination of soil and ground water (Fan, 1988). Recently, its toxicity and carcinogenicity to humans have been concerned. By these reasons, it is highly required to clean contaminated soils and ground water (Chemicals Evaluation and Research Institute, Japan, 2006). Though the aeration method has often been used up to now for cleaning contaminated soils, its high remediation cost is one of the serious problems (Komatsu, 2005). In the present study, the possibility of application of electrochemical treatment method to the *in-situ* decomposition of TCE is examined because *in-situ* remediation is expected to be simple and inexpensive. Electroosmotic flow is the motion of liquid induced by an applied potential across porous materials, capillary tubes, membranes, microchannels, and any other fluid conduits. Because electroosmosis method has the following advantages, it is expected to be used for *in-situ* cleaning of the contaminated soil. Especially, it can be applied to the soils with low water permeability such as clay and silt. Furthermore, the combination of this method with other purification ones is easy (Ninae et al., 2001, 2005; Athmer et al., 1994). Moreover, the decomposition of the pollutant using radicals from various electrochemical reactions produced at the electrode surface is reported (Wang & Lemley, 2002). When water is electrically decomposed by using the stable electrode, oxygen and radical compound are generated in the anode. Although the life time of this radical compound is very short, its strong oxidizing power is expected to be strong enough to decompose persistent substances.

The authors have reported that TCE is decomposed by the electro degradation simultaneously with the electroosmosis (Matsumura et al., 2001). The experiment was conducted in the solution as a basic examination of the electroosmosis method. However, the ascertainment of the effectiveness of this method in the soil has not so far been conducted. In the present study, the experiment in the artificial soil was conducted as a simulation of remediation of contaminated soil.

2. Material & Method

2.1 TCE Decomposition in Solution

One thousand mL of various concentrations (0.05–1.0 mmol/L) of the aqueous solution of potassium nitrate including TCE (4.0 mmol/L) was prepared for 180 minute test at 15 °C in a glass vessel (100 mm × 100 mm × 200 mm). The platinum (anode) and the stainless (cathode) plates were used as the electrodes. The electric power with various currents was applied under the constant current condition and the voltages were read and 3 ml of the solution was sampled from the reactor at regular intervals. The residual TCE was extracted from the solution with 15 ml of n-hexane (5 ml × 3). It is confirmed that all TCE is extracted with this amount of n-hexane in the preliminary experiment. The n-hexane solution was dehydrated with anhydrous sodium sulfate. The concentration of the residual TCE in the n-hexane solution was quantified by a gas chromatograph with an electron capture detector (GC-ECD). The experimental apparatus chart used is shown in Figure 1.

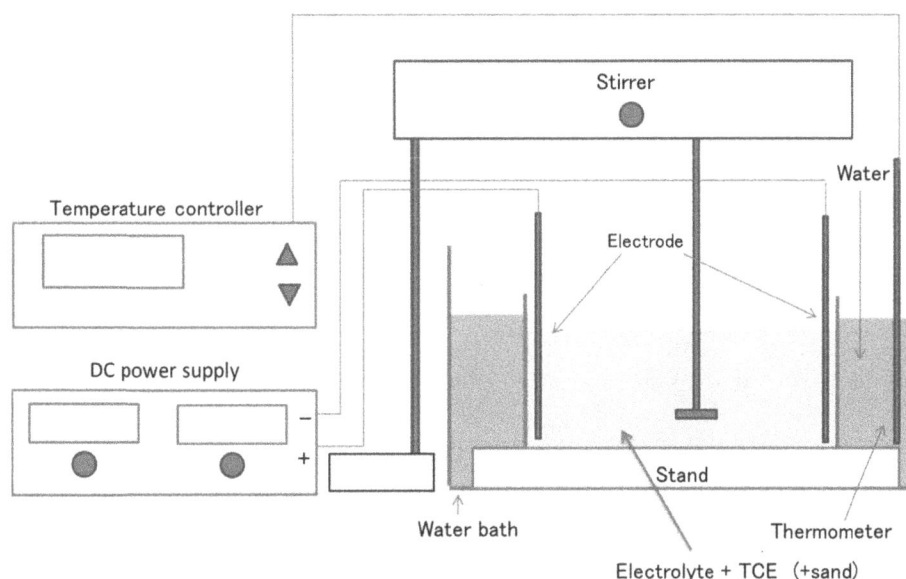

Figure 1. Experimental setup of electro degradation

2.2 TCE Decomposition in Soil

Same vessel as above was used in this experiment. The same vessel as above was packed with 1000 mL of aritificial soil (Silica sand, bed voidage 34.0%; Glass bead, bed voidage 40.4%). Various concentrations (0.05–1.0 mmol/L) of the aqueous solution of potassium nitrate including TCE (4.0 mmol/L) was poured until the void of the soil was saturated in the reactor filled with artificial soil. The experiments were conducted under the condition similar to the above condition for solution only, excepting that soil layer was not stirred. The solution was energized at various constant currents. Three mL of solution in the soil was sampled from three places (the vicinity of anode, the center and the vicinity of cathode) by 10 g at regular intervals. The residual TCE in the soil was extracted by acetonitrile 20 ml and water 10 ml, and the TCE concentration was determined as above.

3. Result and Discussion

3.1 TCE Decomposition in Solution

An example of time variation of residual TCE [wt%] by electrical decomposition is shown in Figure 2. The reaction path through which the hydroxide radical generated from the electro degradation of water decomposes TCE is reported as follows (Yoshida, 2008; Lu et al., 2009)

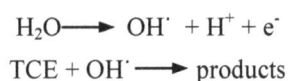

$$H_2O \longrightarrow OH^{\cdot} + H^+ + e^-$$

$$TCE + OH^{\cdot} \longrightarrow products$$

The theoretical value of time for complete decomposition of TCE was calculated assuming the above stoichiometric equation, namely one mol of TCE is decomposed by one Faraday of electricity. Namely it proceeds by the zero-order reaction kinetics. Under the condition of initial TCE concentration of 4.0 mmol/L and current of 100 mA, the time for complete decomposition of TCE was calculated to be about 64 min, using the aqueous volume of 1 L. This calculated value is the time when the added electric quantity reaches 4.0 mF because 1 F is necessary to decompose TCE of 1 mol. As shown from Figure 2, the 60% of TCE remains at the reaction time of 60 minutes. It is suggested that the ratio of TCE reacted to the applied electron was less than unity under the actual condition, which is caused by that the electro degradation of water is a primary reaction and that life time of hydroxide radicals is short.

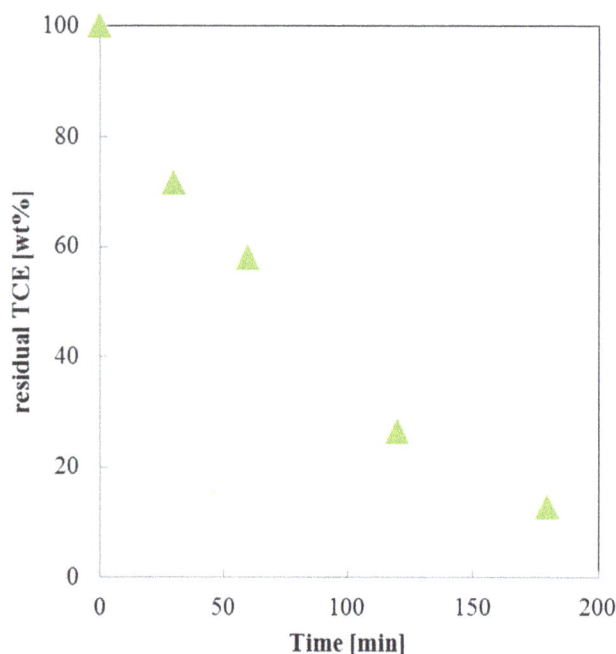

Figure 2. An example of time variation of residual TCE by electrical decomposition in water (current, 100 mA: KNO$_3$ Concentration, 0.10 M)

Hence, this reaction thereafter was calculated as a first-order reaction which well described the progress of the TCE decomposition reaction. First-order reaction rate constant, k, was obtained from the slope of the regression line on semi-log plot of residual TCE concentration, C_{TCE}, vs. reaction time, t, according to the following equation.

$$\ln(C_{TCE}/C_{TCE,0}) = -kt \qquad\qquad (1)$$

The effect of the potassium hydroxide concentration on the relationship between the first order rate constant and applied current is shown in Figure 3 under the constant current. The resulted voltage and power are also calculated and the observed rate constant data are also plotted against them respectively in Figures 4 and 5.

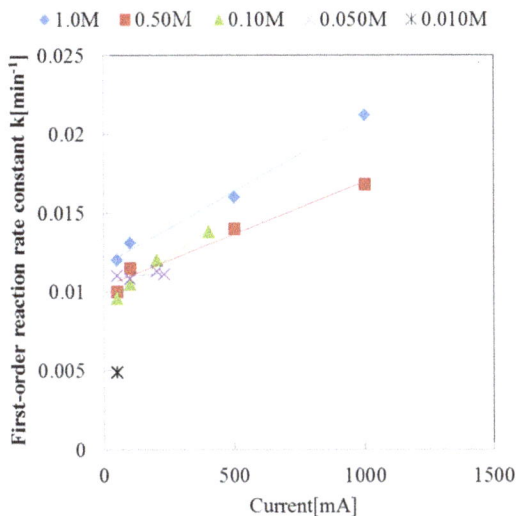

Figure 3. Electro degradation of TCE in water (Constant current condition)

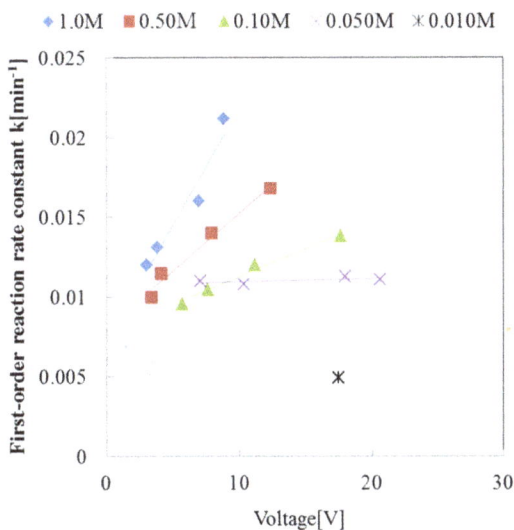

Figure 4. Electro degradation of TCE in water (Constant current condition shown in Figure 3)

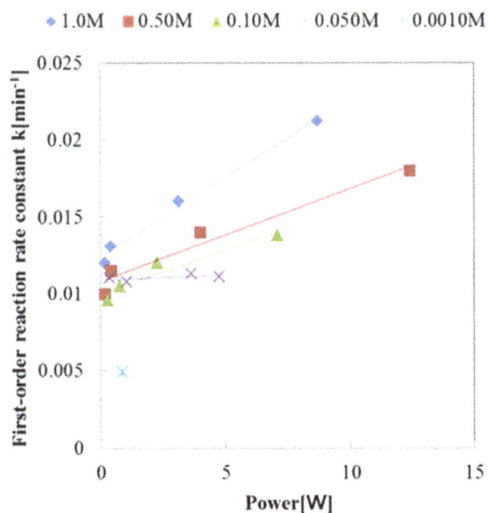

Figure 5. Electro degradation of TCE in water (Constant current condition shown in Figure 3)

As shown from Figures 3–5, the TCE decomposition rate was increased with an increase in electric strength condition and the increased electrolyte concentration increased the rate of reaction. Therefore, it was thought that the radical formation speed on the electrode surface is rate-determining step or the transfer rate of TCE is accerelated by the increased radical concentration.

Next, the effect of the stirring speed on the relation between the concentration of TCE and reaction time was shown in Figure 6. The theoretical values of mass transfer rate of TCE calculated using the Johnson-Huang equation and the Wilke-Chang equation are also shown as straight lines in Figure 6 for comparison. The Johnson-Hung equation for the boundary film mass transfer estimation, the Wilke-Chang equation for diffusion coefficient estimatin, and the differential equation used for theoretical prediction are shown below. In the cause of the calculation of Equation 2, we approximately used 0.1 m for the diameter of reactor, D_T, though our reactor was rectangular and transfer area is vertical while the reactor of Johnson & Huang was cylindrical and transfer area is horizontal and accordingly, the absolute value of mass transfer rate might be different.

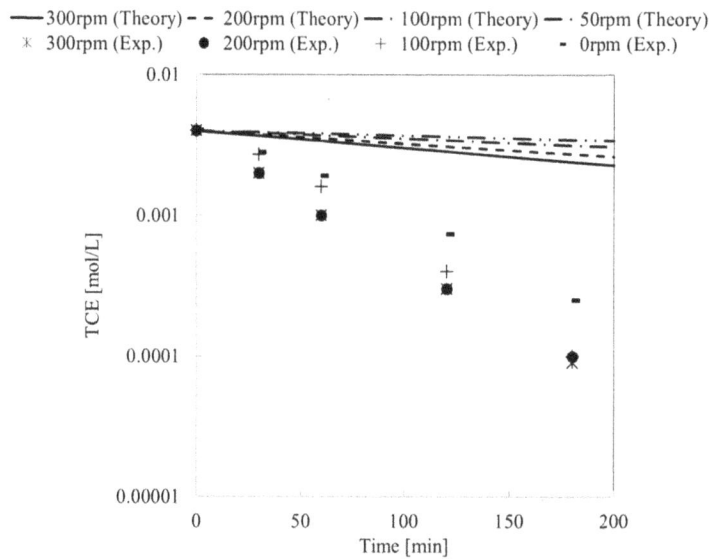

Figure 6. Comparison of experimental results and theoretical values assuming boundary film mass transfer control

< Johnson-Hung equation (Johnson & Huang, 1956) >

$$Sh = 0.0924 \, Re^{0.71} \, Sc^{0.5} \tag{2}$$

< Equation of dimensionless number >

$$\text{Sherwood number: } Sh = \frac{k_L D_T}{D_A} \tag{3}$$

$$\text{Reynolds number: } Re = \frac{d^2 n \, \rho}{\mu} \tag{4}$$

$$\text{Schmid number: } Sc = \frac{\mu}{\rho \, D_A} \tag{5}$$

< Wilke-Chang equation (Wilke & Chang, 1955)>

$$D_A = \frac{7.4 \times 10^{-8} \left(\beta \, M_{H_2O} \right)^{0.5} T}{\mu \, V_m^{0.6}} \tag{6}$$

< Equation of theoretical value >

$$-V \frac{dC_{TCE}}{dt} = k_L A C_{TCE} \tag{7}$$

Or solution

$$\ln \frac{C_{TCE}}{C_{TCE,0}} = -\frac{A}{V}k_L t \qquad (8)$$

$$C_{TCE} = C_{TCE,0} e^{-\frac{Ak_L}{V}t} \qquad (9)$$

k_L:	Mass transfer coefficient, m/s	β:	Association coefficient, -
D_T:	Diameter of reactor, m	M_{H_2O}:	Molecular weight of water, g/mol
D_A:	Diffusion coefficient of TCE, m²/s	V_m:	Molar volume of TCE, cm³/mol (=10⁻⁶m/mol)
d:	Diameter of stirring blade, m	C_{TCE}:	TCE concentration at time t, mol/m³
n:	Rotating speed (in rps), s⁻¹	$C_{TCE,0}$:	TCE initial concentration, mol/m³
ρ:	Density of the solution, kg/m³	A:	Surface area of electrode, m²
μ:	Viscosity of solution, Pa·s (cP = mPa·s, only in Equation (6))	t:	Time, s
		V:	Volume of liquor, m³
T:	Temperature, K		

As shown from Figure 6, the stirring speed did not influence the decomposition rate of TCE under the present experimental condition while the theoretical decomposition rate of TCE increases with an increase in the stirring speed under the condition of film mass transfer control. Therefore, the rate-determining step of the electro degradation of TCE is not the simple transport of TCE in the film. Considering the experimental decomposition rate was much faster than that of the predicted value under the mass transfer control, it is suggested that the diffusion of TCE in the film is accelerated by the electrically radicals and it is the rate controlling step. Consequently, it is thought that the electro degradation rate of TCE was influenced by the electric condition.

3.2 TCE Decomposition in Soil

Effect of voltage on the relationship between reaction time and residual TCE concentration in electro degradation of TCE in the soil (quartz sand) is shown in Figure 7. The slope of the fitting line gives the apparent first-order reaction rate constant. For comparison, the result in solution is also shown in Figure 7. As shown from Figure 7, the decomposition rate in the soil doesn't increase even with increased voltage. In addition, the decomposition rate of TCE in the soil (sand) is slower than that in the solution. It is thought that TCE is not sufficiently supplied to the electrode surface by mass transfer resistance in the soil, though the electro-osmotic flow may be caused in the soil.

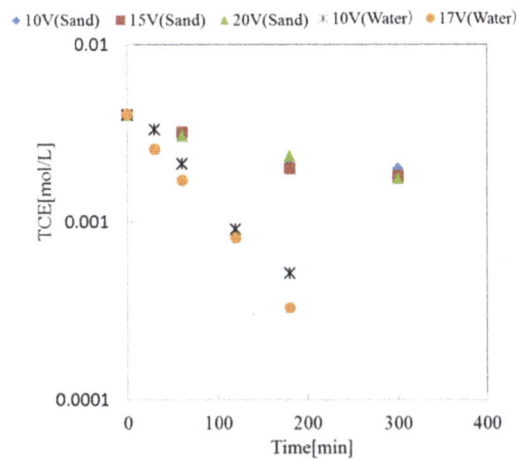

Figure 7. Electro degradation of TCE in soil

Next, effect of particle size in the soil layer on the time variation of residual TCE concentration in electro degradation of TCE is shown in Figure 8. As shown from Figure 8, the TCE decomposition rate is almost equal in the both cases with quartz sand (156 μm) and glass bead (386 μm) used as artificial soil. Considering that the resistance of bulk flow is affected by the particle diameter, while diffusion is mainly affected by the voidage, it turned out that TCE is not transported by bulk flow but is mostly transferred by molecular diffusion in the voidage of the space under the actual experiment condition. The reason why the same results are given for both of soils is explained by the fact that not only the effective diffusion coefficient but also the volume of the solution is proportional to the voidage, leading to the same time variation of concentration of TCE in the soil bed void.

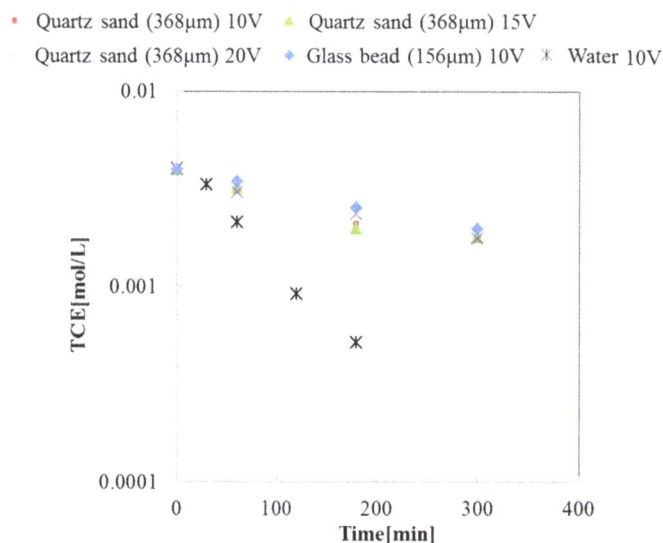

Figure 8. Influence of difference in particle diameter

4. Conclusion

In-situ decomposition of TCE using the electrochemical treatment method in the solution and in the soil was investigated. The following characteristics of the reaction were suggested.

1) The rate-determining step of the electro degradation of TCE is not simple diffusion of TCE in boundary film but is the accelerated by the decomposition of TCE by radicals formed at or near the electrode surface. Therefore, an increase of the electrolyte concentration and an increase in the electric power are effective.

2) In the electro degradation of TCE in the soil, the transportion of TCE to the electrode surface is the rate-determining step, and the TCE decomposition rate is much slower than that in solution. Moreover, it turned out that TCE is not transported by bulk flow but is mostly transfered by molecular diffusion in the soil.

References

Athmer, C. J., Ho, S. V., & Sheridan, P. W. (1994). Laboratory Study of the Movement of Trichloroethylene through Clayey Soils by Electroosmosis. *Proc Annu Meet Air Waste Manag Assoc, 87*(14), 1-16.

Chemicals Evaluation and Research Institute, Japan. (2006). *HAZARD ASSESSMENT REPORT Trichloroethylene*. Tokyo: Chemicals Evaluation and Research Institute, Japan.

Fan, A. M. (1988). Trichloroethylene: water contamination and health risk assessment. In *Reviews of environmental contamination and toxicology* (pp. 55-92). Springer New York.

Johnson, A. I., & Huang, C. J. (1956). Mass transfer studies in an agitated vessel. *AIChE Journal, 2*(3), 412-419. http://dx.doi.org/10.1002/aic.690020322

Komatsu, T., & Møldrup, P. (2005). The emerging role of soil physics in environmental engineering. *Journal of the Japanese Society of Soil Physics, 100*, 5-14.

Lu, X., & Yuan, S. (2009). Electrokinetic removal of chlorinated organic compounds. In K. R. Reddy & C. Cameselle (Eds.), *Electrochemical Remediation Technologies for Polluted Soils, Sediments and Groundwater* (pp. 219-234). Hoboken, New Jersey: John Wiley & Sons, Inc.

http://dx.doi.org/10.1002/9780470523650.ch10

Matsumura, M., & Kojima, T. (2001). Decomposition of trichloroethylene in soil by electric osmosis. *Kagaku Kogaku Ronbunshu, 27*(1), 1-6. http://dx.doi.org/10.1252/kakoronbunshu.27.1

Niinae, M., Aoki, Y., & Aoki, K. (2005). Numerical Modeling for Electrokinetic Soil Processing-Acidification of Soil. *Journal Code: F0195A, 52*(3), 136-144.

Ninae, M., Sugano, T., Aoki, K., & Yasuda, T. (2001). A Study on Contaminant Removal from Soil by Coupled Electric. Hydraulic Gradient. *Resources Processing, 48*(3), 176-183. http://dx.doi.org/10.4144/rpsj1986.48.176

Wang, Q., & Lemley, A. T. (2002). Oxidation of carbaryl in aqueous solution by membrane anodic Fenton treatment. *Journal of Agricultural and Food Chemistry, 50*(8), 2331-2337. http://dx.doi.org/10.1021/jf011434w

Wilke, C. R., & Chang, P. (1955). Correlation of diffusion coefficients in dilute solutions. *AIChE Journal, 1*(2), 264-270. http://dx.doi.org/10.1002/aic.690010222

3

Co-Production of Liquid and Gaseous Fuels from Polyethylene and Polystyrene in a Continuous Sequential Pyrolysis and Catalytic Reforming System

Mochamad Syamsiro[1*], Wu Hu[1*], Shuta Komoto[1], Shuo Cheng[1], Putri Noviasri[1], Pandji Prawisudha[2] & Kunio Yoshikawa[1]

[1] Department of Environmental Science and Technology, Tokyo Institute of Technology, Yokohama, Japan

[2] Department of Mechanical Engineering, Institut Teknologi Bandung, Bandung, Indonesia

* These authors contributed equally.

Correspondence: Mochamad Syamsiro, Department of Environmental Science and Technology, Tokyo Institute of Technology, Japan. E-mail: syamsiro@yahoo.co.id

Abstract

This paper deals with the potential of using sequential pyrolysis and catalytic reforming process in a continuous system for the conversion of polyethylene and polystyrene into liquid and gaseous fuels using the HY-zeolite catalyst for the catalytic reforming of pyrolysis gas generated in a pyrolyzer. The effect of reforming temperature and the weight hourly space velocity on the product yields, liquid and gaseous compositions have been investigated for each feedstock. The experiments were carried out at the pyrolyzer temperature of 450 °C, the reforming temperature of 400, 450, and 500 °C and the weight hourly space velocity of 2, 3, and 4 g-sample g-catalyst^{-1} h^{-1}. The results show that increasing the reforming temperature and decreasing the weight hourly space velocity have resulted in an increase of gaseous and solid products while the liquid product decreased. The maximum oil production for HDPE (70.0wt%) and PS (88.1wt%) were obtained at the pyrolysis temperature of 450 °C, the reforming temperature of 450 °C and the weight hourly space velocity of 4. The C2, C3 and C4+ gases (>75 mol %) were the main components of the gaseous and liquid products for HDPE. In case of PS, the C2 and C3 gases (>65 mol %) were the main components of the gaseous product. The high quality of gaseous products can be used as a fuel either for driving gas engines or for dual-fuel diesel engines.

Keywords: waste plastics, pyrolysis, catalytic reforming, fuels, zeolite

1. Introduction

Plastics are now becoming substantial materials in modern life and have wide range of applications. Plastic consumption has been growing rapidly in the last six decades due to their ability to be simply formed, its light weight together with non-corrosive behavior. The world's annual plastic consumption has increased about 20 times from 5 million tons in 1950s to nearly 100 million tons (United Nations Environment Programme [UNEP], 2009). A significant growth of the plastic consumption has resulted in an increased production of plastic wastes. Thus, plastic wastes have become a major stream in solid waste and caused significant environmental problems for nations worldwide. Disposing of plastic wastes by landfilling is not a suitable option due to slow degradation rates. The use of incineration technology has caused environmental problems since it generates several pollutants to the atmosphere. To minimize the environmental impact and to reduce damages caused by plastic wastes, they must be recycled and recovered. Therefore, alternative methods such as chemical or feedstock recycling which involves pyrolysis of plastics into fuel have been introduced not only for waste reduction but also for fuel production. This method has become very promising technology since plastics have a high calorific value of more than 40 MJ/kg which is similar to those of common liquid fuels such as gasoline, diesel, kerosene, etc.

Pyrolysis or thermal cracking involves the degradation of the polymeric materials by heating in the absence of oxygen. The process is usually conducted at temperatures between 500-800 °C (Aguado, Serrano, Miguel, Castro, & Madrid, 2007). These pyrolytic products can be divided into a gas fraction, a liquid fraction and solid residues (Buekens & Huang, 1998). Plastic wastes can be decomposed as a single feedstock or mixed with other materials such as coal and biomass (Ishaq et al., 2006). Co-pyrolysis of plastic and coal has indicated that there was significant synergistic effect between plastic and coal, especially in the high temperature region (Zhou, Luo, & Huang, 2009). The thermal degradation of plastics may involve three different decomposition pathways (Aguado & Serrano, 1999) : (i) random scission at any point in the polymer backbone leading to the formation of smaller polymeric fragments as primary products, (ii) end-chain scission, where small molecules and long-chain polymeric fragments are formed, (iii) abstraction of functional substituents to form small molecules.

In many cases, several of these pathways occur simultaneously. However, the thermal degradation of plastics has a major drawback such as very broad product range and requirement of high temperature. These facts strongly limit their applicability and especially increase the cost of feedstock recycling for waste plastic treatment (Lin et al., 2010). Catalytic degradation therefore provides a means to address these problems. The use of catalyst is expected to reduce the reaction temperature, to promote decomposition reactions, and to improve the quality of the products.

Both homogeneous and heterogeneous catalysts have been used for studying the catalytic cracking of plastics by many researchers. In general, heterogeneous catalysts are the preferred choice due to their easy separation and recovery from the reacting medium (Aguado, Serrano, & Escola, 2006). A wide variety of heterogeneous catalysts have been tested by researchers such as zeolite, silica alumina, and FCC catalyst. Each catalyst has different structure and composition which affect the properties of the fuel products.

Catalytic degradation of plastic wastes has been investigated extensively by many researchers using zeolite-Y, ZSM-5, mordenite and silica alumina (Seo, Lee & Shin, 2003; Mikulec & Vrbova, 2008; Wang & Wang, 2011). The presence of catalysts promotes the chemical reaction resulting in the selectivity production of specific products with high added value. Direct catalytic cracking has been used widely due to several advantages, mostly in terms of the energy efficiency, the reaction temperature and the residence time. However, direct catalytic cracking of plastic waste suffers from a number of drawbacks which have prevented its commercial success. The first relates to difficulty to recover the catalyst after use, which increases the operational cost. Furthermore, direct contact with plastic wastes will make catalyst deactivate rapidly due to the deposition of carbonaceous matter and the poisoning effect of extraneous elements and impurities such as chlorine, sulfur and nitrogen containing species that may be present in the plastic wastes (Aguado et al., 2007). Therefore, separation of the catalytic reforming reaction from the pyrolysis stage can be applied to overcome these problems. This method has been firstly tested by Bagri and Williams (2002, 2004) for polyethylene and polystyrene using zeolite-Y and ZSM-5 catalysts. The use of other catalysts such as silica alumina and Al-MCM-41 have also been investigated by others (Wang & Wang, 2011; Miguel, Serrano, & Aguado, 2009). Preliminary assessment of plastic wastes valorization by using this method has been studied by Iribarren, Dufour & Serrano (2012). From a combined energy and environmental perspective, the results suggested the suitability of this system for plastic waste valorization. The energy performance of this system was deemed appropriate, based on the calculated cumulative energy demand and net energy ratio values.

The low thermal conductivity and high viscosity of plastics are the major problems for the cracking reactor design. Therefore, the reactor design becomes important parameter in feedstock recycling of plastics. Several reactor systems have been developed and used such as batch/semi batch, fixed bed, fluidized bed, spouted bed and screw kiln. Batch or semi-batch reactors have been used by many researchers because of its simple design and easy operation. However, it has a drawback in the stability of the process especially for large scale applications. Therefore, continuous flow operation is a suitable technique to study the degaradation of plastics because the experimental data are obtained at steady state, that is, at a constant temperature, a constant pressure, and a constant amount of reactor content (Murata, Brebu, & Sakata, 2010).

In this paper, the sequential pyrolysis and catalytic reforming (SPCR) of polymer in a continuous system has been proposed to produce liquid and gaseous fuels over the HY-Zeolite catalyst. The fixed bed reactor was used as the pyrolysis reactor under the atmospheric pressure. Most of previous researches utilizing SPCR process have been done in a batch system. We introduced a continuous system utilizing SPCR process to produce liquid and gaseous fuels. Our proposed system will utilize all of products as fuels including liquid, gaseous and solid products. This novel system will utilize diesel fuel and gaseous products together for fueling a dual-fuel diesel engine to generate electricity. This power can be utilized for supplying the electricity to the plant itself and the excess power will be sent to outside of the plant. The solid products will also be investigated in terms of the energy content to assess the feasibility as a fuel for co-combustion with coal and biomass. The HY-Zeolite catalyst has been used by Bagri and Williams (2002, 2004). However, they studied only the effect of the reforming temperature. In this paper, we investigated not only the effect of the reforming temperature but also the effect of the catalyst loading on the liquid and gaseous products characteristics.

2. Materials and Methods

2.1 Materials

The feedstocks used for these experiments were high density polyethylene (HDPE) and polystyrene (PS) granules manufactured by Tosoh Co. in Japan. The catalyst employed in this study was commercial pelletized HY-Zeolite (CBV 780 CY) obtained from the Zeolyst International. The HY-Zeolite has SiO_2/Al_2O_3 mole ratio of 80, the unit cell size of 24.24 ⑥ and the surface area of 780 m^2/g in the powder form. The diameter of the pellet was 1.6 mm which contains 20% of aluminum oxide.

2.2 Experimental Procedure

A schematic diagram of the experimental apparatus is shown in Figure 1. The apparatus was composed of a feeder, a pyrolyzer, a packed-bed catalytic reformer, a condenser, an oil collector and gas scrubbing bottles. The pyrolyzer and the reformer were made of stainless steel (SUS316) and covered with electric heaters. The pyrolyzer's inner diameter and height are 30 mm and 280 mm, respectively. The reformer's inner diameter and height are 45 mm and 550 mm, respectively. The reaction temperatures in both the pyrolyzer and the reformer were controlled with K-type thermocouples and heaters. A double-tube condenser was installed at the outlet of the reformer to separate gas and liquid products. The gas scrubbing bottles were installed after the condenser and isopropanol was used as the scrubbing absorbent to remove some light tar in the gaseous products.

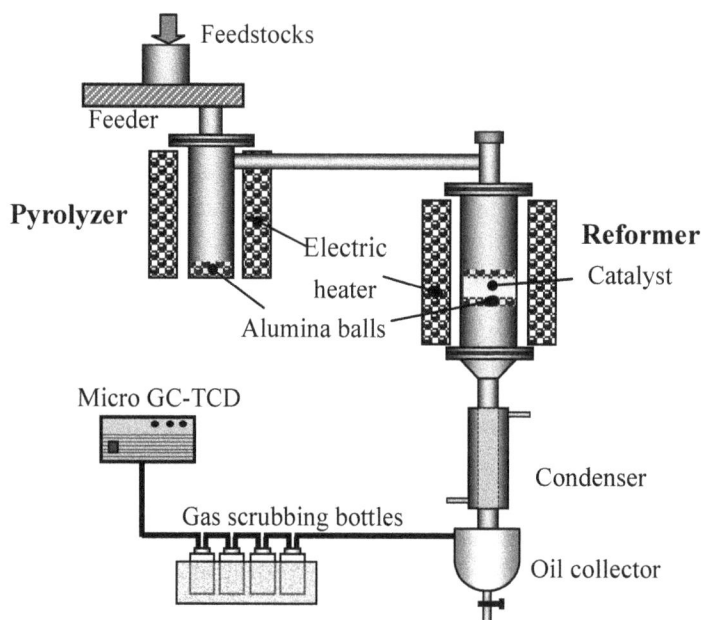

Figure 1. A schematic diagram of the experimental apparatus

In these experiments, after the pyrolyzer and the reformer heated up to the preset temperatures and air in the reactors was replaced with N_2 carrier gas, plastic granules were fed into the pyrolyzer at the feeding rate of 1 g min^{-1}. The feeding time of each experiment was 2 hours. The N_2 carrier gas flow rate was 1 L min^{-1} controlled by using a mass flow controller (Model CR-300, KOJIMA Instruments Inc. Kyoto, Japan). The catalyst was loaded into the reformer with the weight hourly space velocity (WHSV) range of 2, 3 and 4 g-sample g-catalyst^{-1} h^{-1}. The calculation of WHSV is based on the equation below (Park et al., 2010):

$$WHSV = \frac{60 \, x \, G}{W_{cat}}$$

(1)

where G = sample feed rate, g min^{-1}; and W_{cat} = weight of catalyst filled in the reformer, g.

The first experiments were carried out at the pyrolyzer temperature of 450 °C, WHSV of 4 g-sample g-catalyst^{-1} h^{-1} and the reformer temperature of 400, 450, and 500 °C. The second experiments were conducted at the pyrolyzer temperature of 450 °C, the reformer temperature of 450 °C and WHSV of 2, 3, and 4 g-sample g-catalyst^{-1} h^{-1}. The gaseous and liquid products generated in the reformer were separated in the condenser and the liquid product was collected into the oil collector. The gas compositions were measured with a gas chromatograph equipped with a thermal conductivity detector (GC-TCD, Agilent Technologies Inc. USA).

2.3 Analytical Methods

The liquid product were analyzed by a gas chromatograph coupled with a mass spectrometer (GC-MS) (Agilent 6890N GC-MSD 5973N). The purpose of this analysis is to determine carbon atom number distribution and hydrocarbon type of the liquid products. The column was an HP5 (5% Ph-Me-Siloxane) capillary column, 30 m length with 0.25 mm diameter and 0.25 μm film thickness. Helium was used as the carrier gas. The temperature program used was, initial temperature of 30 °C for 5 minutes followed by a heating rate of 2 °C /min to 200 °C and then held at 200 °C for 5 minutes followed by a heating rate of 5 °C min^{-1} to 300 °C and held at 300 °C for 10 minutes.

The composition of gases produced in the experiments was monitored by the GC-TCD every 6 minutes. Gas yields were calculated from Equation. (2)-(4) as follows (Park et al., 2010) :

$$F_{T,out} = F_{N_2,in} \times \frac{C_{N_2,in}}{C_{N_2,out}} \tag{2}$$

$$F_{i,out} = F_{T,out} \times C_{i,out} \tag{3}$$

$$Y_i = \frac{F_{i,out}}{G} \times \frac{1}{22.4} \tag{4}$$

where $C_{i,out}$ = gas i concentration in the outlet gas; $C_{N2,in}$ = N$_2$ concentration in the carrier gas; $C_{N2,out}$ = N$_2$ concentration in the outlet gas; $F_{i,out}$ = gas flow rate of i at the outlet, Nl min^{-1}; $F_{N2,in}$ = carrier gas flow rate (at the inlet), Nl min^{-1}; and $F_{T,out}$ = total gas flow rate at the outlet, Nl min^{-1}.

After finishing the experiments, a small amount of solid residue was remained in the reactor. The coke formation also occurred in the catalyst. Different weight of the catalyst before and after the experiment was defined as coke. The weight of solid residue was calculated by the difference between the total feedstock weight and liquid, gaseous and coke weights.

The proximate and ultimate analysis were conducted for solid residue samples. The higher heating value (HHV) has been calculated using a modified Dulong`s formula as a function of the carbon, hydrogen, oxygen and nitrogen contents as follows (Demirbas, 2010) :

$$\text{HHV (MJ/kg)} = 0.335 \text{ C} + 1.423 \text{ H} - 0.154 \text{ O} - 0.145 \text{ N} \tag{5}$$

where C is carbon content (wt.%), H is hydrogen content (wt.%), O is oxygen content (wt.%), and N is nitrogen content (wt.%).

3. Results and Discussions

The results obtained from the experimental investigation on the pyrolysis and catalytic reforming of HDPE and PS are presented and discussed in this section. The results focused on the effect of the reforming temperature and WHSV on the product yields, oil characteristics and gas composition for each feedstock.

3.1 Pyrolysis and Catalytic Reforming of HDPE

3.1.1 Effect of the Reforming Temperature

The effect of the reforming temperature on the product yields from the pyrolysis and catalytic reforming of HDPE is shown in Figure 2. It can be seen that the increase of the reforming temperature increased the yield of gaseous and coke products, whereas the yield of liquid products was decreased. The higher temperature led to the enhancement of the activity of the HY-zeolite catalyst and then cracked some relatively large-molecular liquid products into small molecular gaseous products. The increase of the coke was mainly originated from the following reasons; on the one hand, the coking reaction was easier to occur on the surface of the HY-zeolite at a higher reforming temperature (Neves, Botelho, Machado, & Rebelo, 2006). On the other hand, the HY-zeolite catalyst has a relatively large pore size and a large supercage with its crystallite, resulting in the formation and accumulation of coke on the internal and external surfaces of the HY-zeolite catalyst. The coke formation may not only prohibit heat transfer and cause operating problems in the reactor, but also lower the number of active sites and the surface area of the catalyst and then lead to the increase of the operation cost (Al-Khattaf, 2002; Neves, Botelho, Machado, & Rebelo, 2007). Therefore, from the viewpoint of industrial application, it is a great importance to optimize the reaction conditions and clarify the deactivation behaviors of the catalyst.

Furthermore, the solid wax was produced in the condenser at the reforming temperature of 400 °C, which means that, at this temperature, the activity of the catalyst was not high enough and could not convert all of the pyrolysis gas into gaseous and liquid products. On the contrary, no wax formation was observed at the reforming temperature of 450 and 500 °C. In addition, the increase of the reforming temperature from 450 to 500 °C resulted in the increase of gaseous products and the decrease of liquid products.

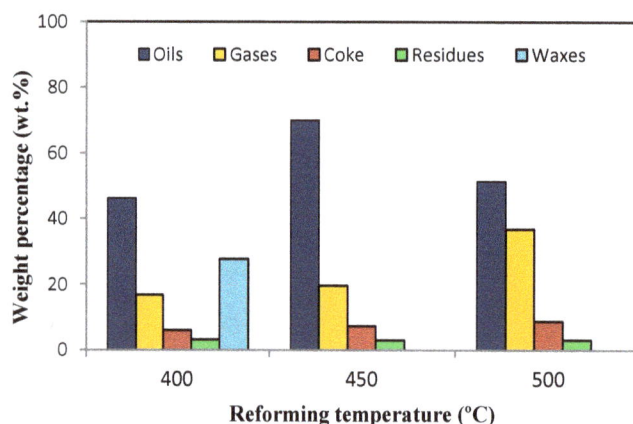

Figure 2. Effect of the reforming temperature on the product yields of HDPE at the pyrolysis temperature of 450 °C and WHSV of 4 g-sample g-catalyst^{-1}h^{-1}

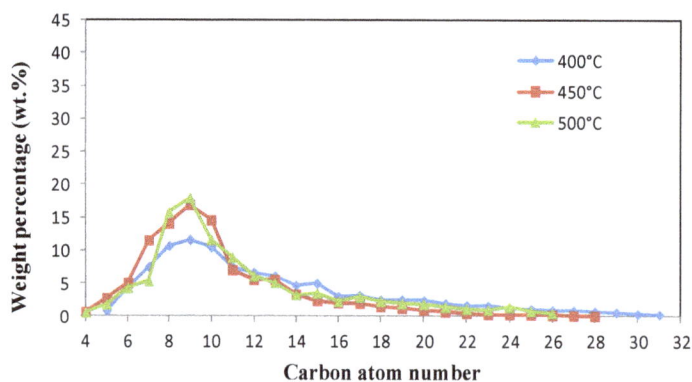

Figure 3. Effect of the reforming temperature on the carbon atom number distribution of liquid products of HDPE at the pyrolysis temperature of 450 °C and WHSV of 4 g-sample g-catalyst^{-1}h^{-1}

The liquid hydrocarbon products generated from the SPCR process were characterized by their carbon atom number distribution and hydrocarbon types which will determine their potential application as a refinery feedstock and fuel. Figure 3 represents the carbon atom number distribution of the liquid products of HDPE produced by the SPCR process at the pyrolysis temperature of 450 °C, WHSV of 4 g-sample g-catalyst^{-1}h^{-1} and at different reforming temperatures from 400 °C to 500 °C. It was notable that the gasoline like components (C_5-C_{12}) and kerosene like components (C_8-C_{16}) were the major components of liquid products. It was due to the relatively moderate acidity and a large pore size of the HY-zeolite catalyst as mentioned previously as well as the type and structure of polyethylene. In addition, the increase of the reforming temperature from 400 °C to 450 °C led to a marked increase in the proportion of light oil (C_5-C_{12}), as well as an obvious reduction in the amount of heavy oil hydrocarbons (>C_{20}), which was attributed to the fact that the higher reforming temperature enhanced the activity of the HY-zeolite catalyst and then led to intense catalytic cracking of the pyrolysis gas into smaller hydrocarbons (Aguado et al., 2007). However, when the reforming temperature increased from 450 °C to 500 °C, the change of the carbon atom number distribution of the liquid products was not significant. It could be the consequence of the coking on the catalyst surfaces at a higher temperature. The formation of coke influenced catalyst activity by covering some of the active sites and blocking the channels which could make the inner active sites inaccessible for the reactant molecules (Miskolczi, Bartha, Deak, Jover, & Kallo, 2004 ; Uemichi, Hattori, Itoh, Nakamura, & Sugioka, 1998).

The effect of the reforming temperature on the distribution of the hydrocarbon types of liquid products from HDPE pyrolysis can be seen in Figure 4. It is well known that the thermal degradation of HDPE has been assigned to the random scission reaction which led to the formation of a large number of n-paraffin hydrocarbon species. The C-C bond is the weakest in the HDPE structure and during the degradation process, the stabilization of the resultant radicals after the chain scission leads to the formation of carbon double bonds in the structure in addition to n-paraffin (Williams & Slanery, 2007). Consequently, the major compositions of the liquid products of HDPE thermal degradation were n-paraffin and olefins. However, in the SPCR process, because of the reforming reaction process, it was found that there was a large number of iso-paraffin, naphthene and aromatic existed in the liquid products of HDPE. The reason for this could be that most zeolites including the HY-zeolite show an excellent catalytic effect on the cracking, isomerization and aromatization due to an acidic property and a micropore crystalline structure. The formation of aromatic compounds was related to both Brönsted and the Lewis sites on the catalysts, but reaction is probably more favorable on the Brönsted sites (Seo et al., 2003). In addition, during the aromatization reaction, a considerable number of hydrogen atoms are abstracted which subsequently accumulate on the catalyst surface, and then consumed in the hydrogenation of olefins. Moreover, the HY-zeolite adsorbed polar molecules strongly, which may also lead to a possible explanation for the increase of iso-paraffins at the expense of olefins (Marcilla, Beltran & Navarro, 2009 ; Manos & Garforth, 2000). Therefore, a lot of aromatics and iso-paraffins at the expense of the n-paraffins and olefins existed in the final liquid products.

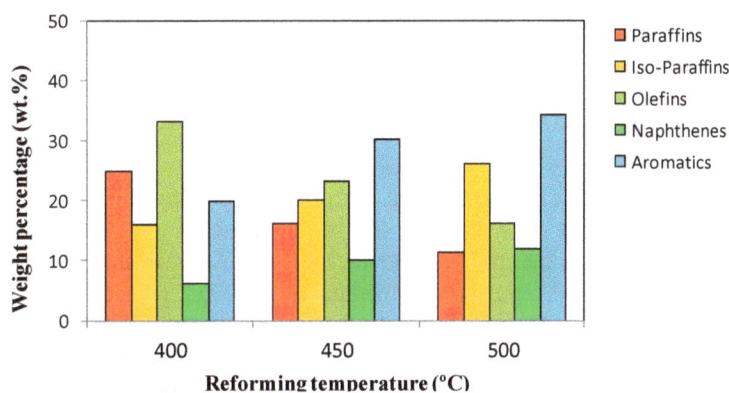

Figure 4. Effect of the reforming temperature on PIONA distribution of HDPE liquid products at the pyrolysis temperature of 450 °C and WHSV of 4 g-sample g-catalyst^{-1}h^{-1}

Futhermore, Figure 4 also indicated that by the increase of the reforming temperature, the proportion of aromatics, iso-paraffins and naphthenes increased while the percentage of paraffins and olefins decreased. The reason is that, as mentioned above, the high temperature will increase the activity of the catalyst and lead to intense catalytic reforming. The reactions such as saturation, isomerization, cyclization, hydrogen transfer and the coking were at high severity (Murata, Brebu, & Sakata, 2009).

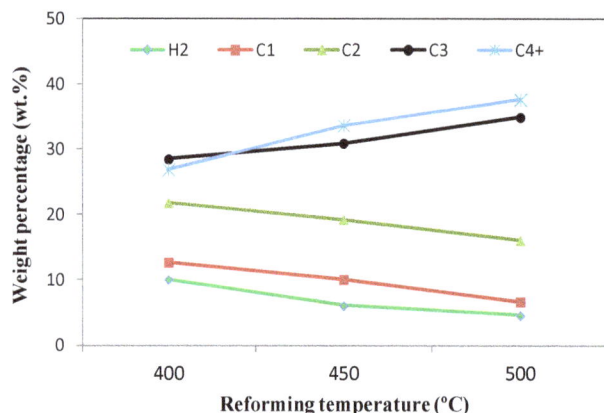

Figure 5. Effect of the reforming temperature on the gaseous product composition of HDPE at the pyrolysis temperature of 450 °C and WHSV of 4 g-sample g-catalyst^{-1}h^{-1}

The gaseous product compositions of HDPE pyrolysis as a function of the reforming temperature is shown in Figure 5. The reforming temperature significantly affected the gaseous composition of the products. It was illustrated from Figure 5 that with the increase of the reforming temperature, there were significant changes of the gaseous compositions of the products. As the reforming temperature increase, the mol percentage of H_2, C_1 and C_2 gases were decreased whereas C_3 and C_{4+} hydrocarbons were increased. The increase of C_3 and C_{4+} hydrocarbons were resulted from the conversion of liquid products into gaseous products. In addition, more than 80 mol% of the final gaseous products was C_2, C_3, C_{4+} hydrocarbons; which originated from the relatively moderate acidity and large pore size of the HY-zeolite catalyst (Miskolczi et al., 2004 ; Audisio & Bertini, 1990 ; Chumbhale et al., 2005). In large scale application, the high quality of gaseous product can be used as a fuel either for driving gas engines or for dual-fuel diesel engines. It can also be used as a heating source for the pyrolysis reactor.

3.1.2 Effect of WHSV

In a continous flow reactor, another parameter affecting the products distribution is the weight hourly space velocity (WHSV). Therefore, the effect of WHSV on the products yields, as well as the composition and physicochemical properties of gaseous and liquid products of HDPE was evaluated in this section at the pyrolysis temperature of 450 °C and the reforming temperature of 450 °C, respectively. It can be seen from Figure 6 that when both the pyrolysis temperature and the reforming temperature were fixed at 450 °C, with the increase of WHSV, the fraction of gaseous products decreased whereas the liquid products increased. This is due to the fact that increasing WHSV is equal to reducing the amount of catalysts and shortening the contact time. In addition, it was found that the effect of increasing the amount of catalyst (decreasing the WHSV) on the product yields have the similar trend with that of increasing the reforming temperature.

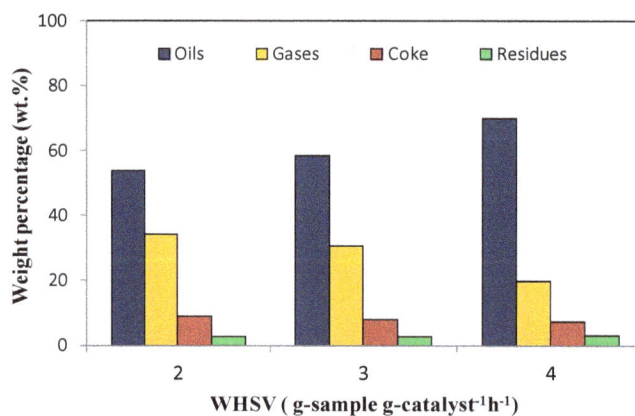

Figure 6. Effect of WHSV on the product yields of HDPE at the pyrolysis temperature of 450 °C and the reforming temperature of 450 °C

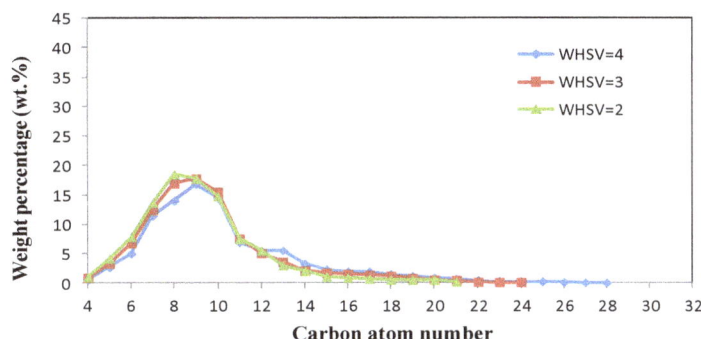

Figure 7. Effect of WHSV on the carbon atom number distribution of liquid products of HDPE at the pyrolysis temperature of 450 °C and the reforming temperature of 450 °C

The effect of WHSV on the carbon atom number distribution of HDPE liquid products was investigated at the pyrolysis temperature of 450 °C and the reforming temperature of 450 °C. Figure 7 indicates that the decrease of WHSV (increase of the catalyst loading) resulted in the increase of the amount of light oil. Therefore, it was found that the effect of decreasing WHSV on the product yields of HDPE has similar trend with that of increasing the reforming temperature. As mentioned above, these results were due to the increase of the catalyst loading and the contact time, which could enhance the cracking reaction of the pyrolysis gas from relatively large-molecule hydrocarbons into small-molecule hydrocarbons. It was also observed from Figure 7 that the increase of WHSV increased the amount of heavy oil. However, light oil (C_5-C_{15}) consisted of gasoline and kerosene fractions was the major component of liquid products, which is caused by the relatively moderate acidity and large pore size of the HY-zeolite (Luo, Suto, Yasu, & Kato, 2000).

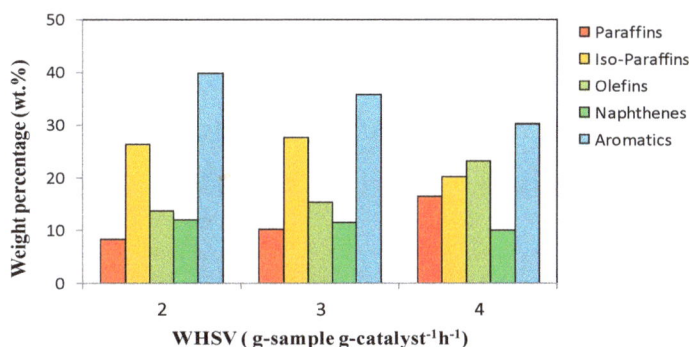

Figure 8. Effect of WHSV on PIONA distribution of HDPE liquid products at the pyrolysis temperature of 450 °C and the reforming temperature of 450 °C

Figure 8 shows the effect of WHSV on the hydrocarbon type distribution of HDPE liquid products. The olefins, iso-paraffins and aromatics were the main type of liquid products. As discussed previously, on the one hand, the HY-zeolite catalyst could provide a large amount of acidic sites. On the other hand, because of its special pore size, the HY-zeolite catalyst has favorable shape selectivity for aromatic formation than non-zeolite catalyst, some intermediate carbenium ion formed by acidic zeolite would choose a pathway to aromatic formation, and some will be left over as olefin. Moreover, Figure 8 also indicates that when both the pyrolysis temperature and the reforming temperature were fixed at 450 °C, the decrease of WHSV led to the increase of the percentage of iso-paraffins, naphthene and aromatic at the expense of olefins and paraffins. This is due to the fact that a lower WHSV means larger amount of catalyst and longer contact time, which could promote the overcracking and secondary reaction such as aromatization, isomerization, etc. It can also be seen from Figure 8 that the increase of the reforming temperature and the decrease of WHSV have the similar trend on the hydrocarbon type distribution of HDPE liquid products (Luo et al., 2000).

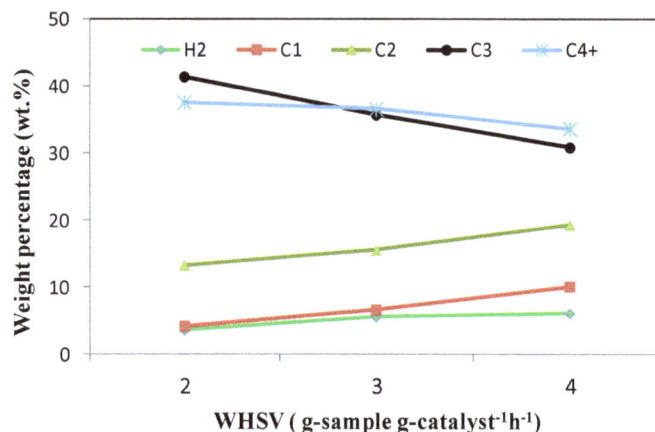

Figure 9. Effect of WHSV on the gaseous product compositions of HDPE at the pyrolysis temperature of 450 °C and the reforming temperature of 450 °C

The effect of WHSV on the composition of gaseous products from HDPE pyrolysis was also investigated at the pyrolysis temperature of 450 °C and the reforming temperature of 450 °C. It is well known that the pore size is important for determining the size selectivity of reactants and products, which can enter and leave the active sites of the catalyst (Uemichi et al., 1998). When compared with other zeolite catalyst, such as HZSM-5, the HY-zeolite catalyst has a relatively large pore size, which may allow a little larger molecular gaseous products to leave the HY-zeolite catalyst. The results shown in Figure 9 illustrates that with the decrease of WHSV, the proportion of H_2, C_1 and C_2 gases decreased while the percentage of C_3 and C_{4+} gases increased, which proves the fact that the HY-zeolite catalyst has special selectivity for C_3 and C_{4+} gases. Therefore, the major gaseous products of the catalytic reforming of pyrolysis gas were C_3 and C_{4+} gases. When increasing the catalyst loading (lowering WHSV), the fraction of gaseous product increased and then the percentage of C_3 and C_{4+} gases increased at the expense of the H_2, C_1 and C_2 gases.

3.2 Pyrolysis and Catalytic Reforming of PS

3.2.1 Effect of the Reforming Temperature

Figure 10 shows the effect of the reforming temperature on the product yields from the pyrolysis and catalytic reforming of PS. It can be seen that PS produced higher liquid products and lower gaseous products compared with HDPE. The quantity of gaseous products was very few to serve as a fuel gas either for engine or for the SPCR process. It was related to the marked differences of molecular structure of HDPE and PS. Therefore, the degradation mechanism and thermal degradation products in the first thermal pyrolysis step are significantly different. The thermal degradation of HDPE consists of free radical formation and hydrogen abstraction steps whereas the thermal degradation of PS is a radical chain process including initiation, transfer and termination steps (Kiran, Ekinci, & Snape, 2000). The main composition of pyrolysis gas of HDPE flowing out from the pyrolyzer was wax while the major pyrolysis products of PS were stable aromatic components as liquid phase such as styrene, which have stable benzene ring structure and difficult to be converted into small molecular gaseous products (Pinto, Costa, Gulyurtlu, & Cabrita, 1999; Liu & Qian, 2000). However, these aromatic hydrocarbons were readily converted into coke existing on the internal and external surface of catalyst to make the catalyst deactivate.

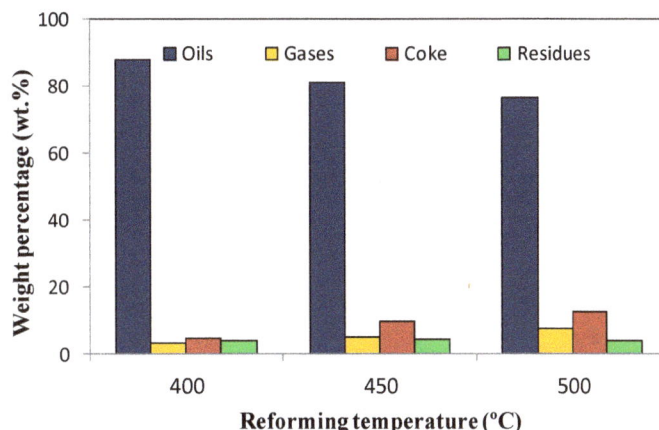

Figure 10. Effect of the reforming temperature on the product yields of PS at the pyrolysis temperature of 450 °C and WHSV of 4 g-sample g-catalyst^{-1}h^{-1}

The effect of the reforming temperature on the carbon atom number distribution from the pyrolysis and catalytic reforming of PS is shown in Figure 11. It indicates that when the pyrolysis temperature and WHSV were fixed at 450 °C and 4 g-sample g-catalyst^{-1}h^{-1}, respectively, the increase of the reforming temperature significantly affected the carbon atom number distribution of the liquid products of PS derived from the SPCR process. In addition, the C_6-C_8 and C_{14}-C_{16} were the main compositions (80wt%) of liquid products derived from PS decomposition, which means the liquid products composition of PS was simpler than that of HDPE. Furthermore, as the increase of the reforming temperature from 400 °C to 450 °C, the proportion of C_6 (benzene) and C_7 (toluene) increased while the proportion of C_8 (styrene monomer and ethylbenzene) and C_{14}-C_{16} (styrene dimer and polycyclic aromatic hydrocarbon) decreased. This is due to that a higher temperature would enhance the activity of the HY-zeolite and then improve the oligomerization and the hydrogen transfer reaction, which gave increase of the formation of monoaromatic hydrocarbon with a smaller molecular weight (Zhang et al., 1995; Onwudili, Insura, & Williams, 2009). However, when the reforming temperature increased from 450 °C to 500 °C, the changes of the carbon atom number distribution of the liquid products of PS was not significant because of the coking formation on the catalyst surfaces at a higher temperature. As mentioned previously, aromatic species have a greater predisposition to be involved in pathways to coke formation because of their ability to easily involve themselves in hydrogen transfer and cyclisation reactions. The analysis of pyrolysis oils derived from polystyrene have been shown to be very high in aromatic compounds and the composition is dominated by the presence of styrene (Williams & Bagri, 2004), consequently, leading to high char formation on the external and/or internal acid sites and then resulted in marked catalyst deactivations, compared with that of HDPE. Therefore, at the reforming temperature of 500 °C, the catalytic performance of the HY-zeolite is not as efficient as the reforming temperature of 450 °C.

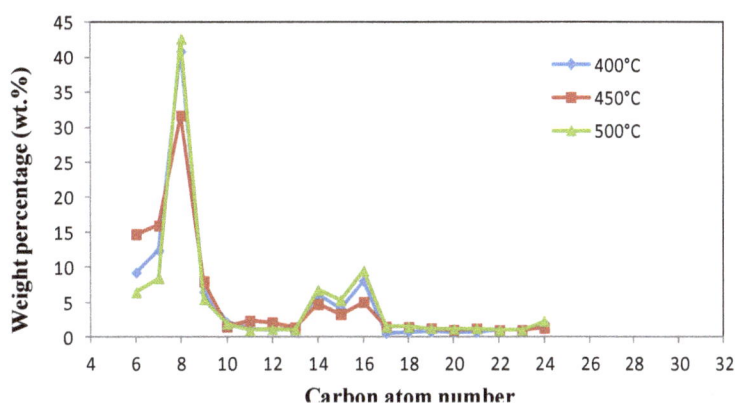

Figure 11. Effect of the reforming temperature on the carbon atom number distribution of liquid products of PS at the pyrolysis temperature of 450 °C and WHSV of 4 g-sample g-catalyst^{-1}h^{-1}

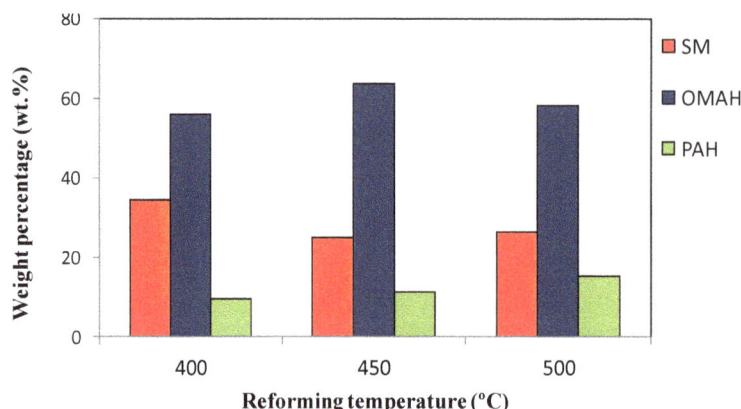

Figure 12. Effect of the reforming temperature on SM, OMAH and PAH distribution of PS liquid products at the pyrolysis temperature of 450 °C and WHSV of 4 g-sample g-catalyst^{-1}h^{-1}(SM: styrene monomer; OMAH: the other monocyclic aromatic hydrocarbon except styrene monomer; PAH: polycyclic aromatic hydrocarbon)

Figure 12 shows the effect of the reforming temperature on the hydrocarbon type distribution of liquid products from PS pyrolysis. It can be seen that when the pyrolysis temperature and WHSV were fixed at 450 °C and 4 g-sample g-catalyst^{-1}h^{-1}, respectively, with the increase of the reforming temperature, the distribution of styrene monomer (SM), the other monocyclic aromatic hydrocarbon (MAH) and polycyclic aromatic hydrocabon (PAH) had an obvious change, which means that the effect of the reforming temperature on the liquid products of PS was significant.

In contrast to HDPE, PS can be thermally depolymerized at a relatively low temperature. The thermal degradation of PS is a radical chain process including initiation, transfer and termination steps in the open reaction system, which led to obtain the styrene monomer with high selectivity. Styrene has a wide range of application in the chemical industry, e.g., in the manufacturing of plastics, synthetic rubber, resins, and insulators. Williams and Williams (1999) have reported that uncatalyzed pyrolysis of polystyrene has been shown to produce 83wt% conversion to a low viscosity oil which consisted mainly of styrene and a gas yield and char yield of less than 5wt% each.

Figure 12 also indicates that there were a large number of ethylbenzene, benzene, toluene, m-xylene, methyl styrene, indan, ethyl methylbenzene and methylbenzene (monocyclic aromatic hydrocarbon), as well as naphthalene, 2-methylnaphthalene, dimethylnaphthalene, phenanthrene, methylnaphthalene and pyrene (polycyclic aromatic hydrocarbon) existed in liquid products of PS. This is mainly due to the presence of relatively moderate acid sites of the HY-zeolites which have the potential to reduce the activation energy of the C-C bond and to hydrogenate the pyrolysis gas into all kinds of aromatic hydrocarbons. As can be seen in Figure 12, considerable amounts of monocyclic aromatic hydrocarbons, such as benzene and ethylbenzene, were formed by the catalytic reforming process. These products are attributed to the further cracking and hydrogenation of styrene yielded from the pyrolyzer, which resulted in a decrease of the faction of styrene in the final oil products. Styrene monomer and polycyclic aromatic hydrocarbons were still main components in the final liquid products of PS, and they might be formed by β-scission of the C-C bond in the polystyrene main chain. Moreover, the increase of the amount of PAH could be related to an increase of intramolecular hydrocarbon transfer at a higher reforming temperature (Zhang et al., 1995).

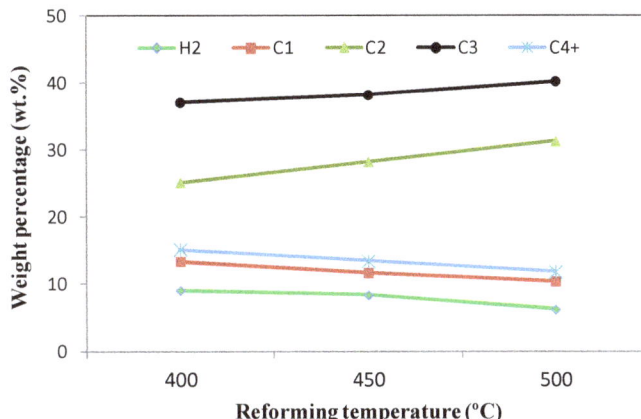

Figure 13. The effect of the reforming temperature on the gaseous product composition of PS at the pyrolysis temperature of 450 °C and WHSV of 4 g-sample g-catalyst^{-1}h^{-1}

The gaseous product composition of PS pyrolysis as a function of the reforming temperature is shown in Figure 13. It can be seen that C_2 and C_3 gases (>65 mol %) were the main components of gaseous products, which originated from the relatively moderate acidity and large pore size of the HY-zeolite catalyst (Miskolczi et al., 2004). As the reforming temperature increase, the mol percentage of H_2, C_1 and C_{4+} gases were decreased whereas C_2, C_3 and hydrocarbons were increased.

3.2.2 Effect of WHSV

Figure 14 illustrates the effect of WHSV on the product yields of PS pyrolysis in SPCR process. It can be seen that the faction of liquid products of PS was higher than that of HDPE while the proportion of gaseous products of PS was much lower than that of gaseous products of HDPE, which means that although increasing the amount of catalysts is equal to increasing the contact time and then improved the overcracking reactions of pyrolysates produced from the pyrolyzer, it was still difficult to reduce the activation energy to break the stable benzene ring structure of the aromatic hydrocarbons of the pyrolysis gas into samller molecular gaseous products (Murata et al., 2009). Furthermore, when compared with that of HDPE, the higher concentraion of aromatic hydrocarbons in the pyrolysis gas of PS also led to more formation of coke on the internal and external surfaces of the HY-zeolite, which led to the increase of the proportion of the solid products. It was mainly due to the fact that, as mentioned previously, the aromatic hydrocarbons especially unsaturated and polyaromatic compounds such as styrene monomer and indan and naphthalene derivatives were formed as major products in the degradation of polystyrene and then they were readily converted into coke (Uemichi et al., 1998).

The carbon atom number distribution of PS as a function of WHSV is illustrated in Figure 15. The carbon atom number distribution of PS was dominated by C_6-C_8 with the weight fraction of more than 70%. This fraction was mainly composed of benzene, methylbenzene, ethylbenzene and styrene, which are valuable chemical feedstock and fuel used in our daily life and modern industry. In addition, there are also certain amount of C_{14}-C_{16} aromatic hydrocarbons existed in the PS oil, which were mainly consisted of some potentially harmful polycyclic aromatic hydrocarbons (Lee, Yoon, & Park, 2002). Therefore, if the oil derived from PS would be used as combustion fuel oil, it should remove these harmful polycyclic aromatic hydrocarbons in advance. Consequently, it is preferable to recycle the PS oil as chemical crude materials rather than as fuel oil, compared with that of HDPE (Joo & Guin, 1997).

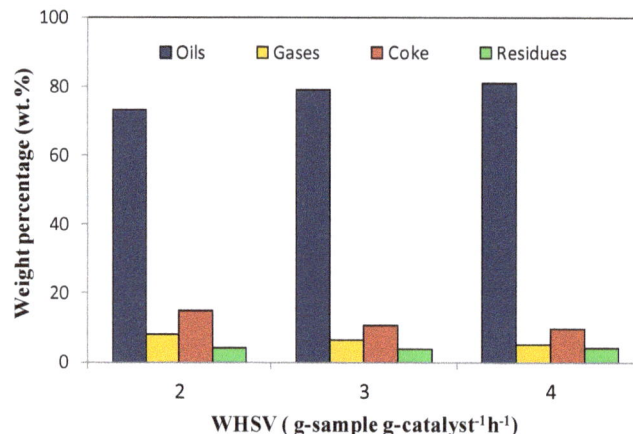

Figure 14. Effect of WHSV on the product phase distribution of PS at the pyrolysis temperature of 450 °C and the reforming temperature of 450 °C

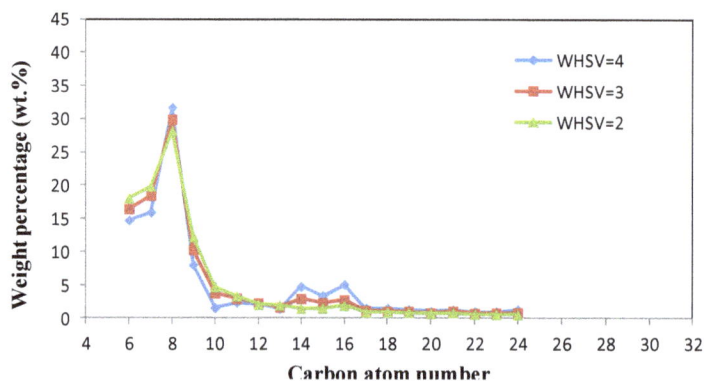

Figure 15. Effect of WHSV on the carbon atom number distribution of liquid products of PS at the pyrolysis temperature of 450 °C and the reforming temperature of 450 °C

Figure 16 shows the effect of WHSV on the hydrocarbon types of PS liquid products at both the pyrolysis temperature and the reforming temperature of 450 °C. It can be observed that monocyclic aromatic hydrocarbons including the styrene monomer were the major component of liquid products of PS. As discussed above, the possible reason was that the presence of the HY-zeolite would improve the hydrogenation reaction, the oligomerization reaction and the β-scission reaction, which would convert the styrene monomer produced from the fast pyrolysis of PS in the pyrolyzer into benzene, methylbenzene and ethylbenzene (monocyclic aromatic hydrocarbon), as well as naphthalene, 2-methylnaphthalene, phenanthrene (polycyclic aromatic hydrocarbon), etc. In addition, the increase of the catalyst loading (decreasing WHSV) enhanced above conversion trend and then led to the increase of polycyclic aromatic hydrocarbon and other monocyclic aromatic hydrocarbon at the expense of styrene monomer. The derived PS oil could be mixed with other kinds of derived oil to improve their RON number, because of the high concentration of aromatic hydrocarbon of PS liquid products (Serrano, Aguado, & Escola, 2000).

The effect of WHSV on the composition of PS gaseous products was also investigated at the pyrolysis temperature of 450 °C and the reforming temperature of 450 °C. As mentioned above, decreasing WHSV was nearly equal to increasing the amount of catalysts and increasing the contact time, which improved the overcracking of liquid products into more small molecular gaseous products. The results shown in Figure 17 illustates that by the decrease of WHSV, the proportion of C_2 and C_3 of PS increased, which means that the HY-zeolite catalyst had special selectivity of C_2 and C_3 gases for PS. This result was consistent with some previously published reports. For instance, Puente and Sedran (1998) also reported a marked increase in C_2 and C_3 gases for the catalyzed pyrolysis of polystyrene compared to the thermal pyrolysis. Williams and Bagri (2004) also reported that C_2 and C_3 gases were the major composition of the gaseous products when employing the HY-zeolite as the catalyst to recycle the polystyrene into valuable fuel and chemical feedstock in a batch system.

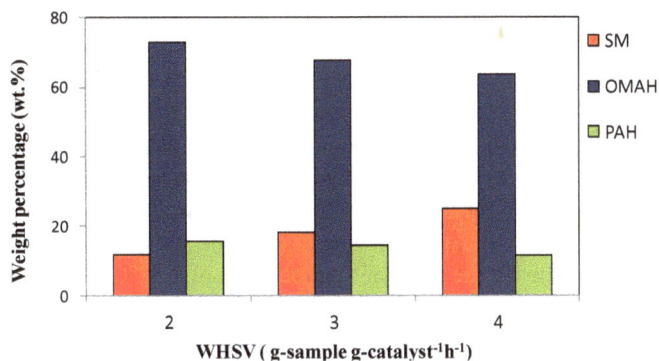

Figure 16. Effect of WHSV on SM, OMAH and PAH distribution of PS liquid products at the pyrolysis temperature of 450 °C and the reforming temperature of 450 °C

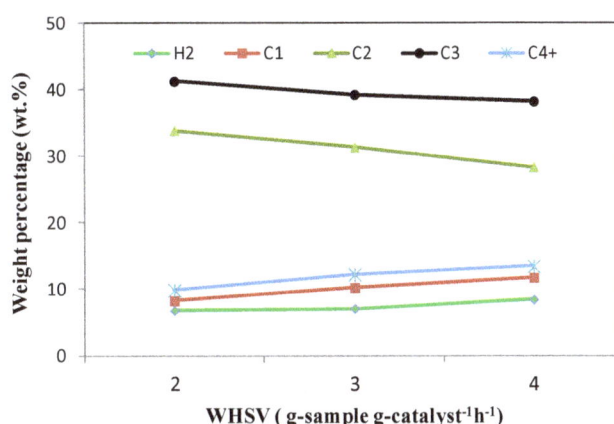

Figure 17. Effect of WHSV on the gaseous product compositions of PS at the pyrolysis temperature of 450 °C and the reforming temperature of 450 °C

3.3 Solid Residues

The proximate analysis and ultimate analysis of the solid residues produced from the pyrolysis of HDPE and PS are shown in Table 1. It can be seen that the solid residue from HDPE has a higher ash content than that of PS. This might be due to the different additives which normally used in plastics materials. The moisture content of HDPE solid residue was also higher than that of PS solid residue. The higher content of ash in HDPE solid residue will absorb more moisture constituent from the atmosphere. The results also show that the solid residues produced from plastics pyrolysis have higher heating value than those of biomass. The higher heating value (HHV) of PS solid residue was higher than that of HDPE residue due to less content of ash in PS solid residue. Therefore, they can be used as a fuel either for blending with biomass and coal or for single fuel.

Table 1. Proximate and ultimate analysis of solid residues (wt.%)

Solid residues	Proximate analysis			Ultimate analysis						HHV** (MJ/kg)
	Moisture	Volatile matter	Fixed carbon	Ash	C	H	N	S	O*	
HDPE	3.09	19.14	57.99	19.78	65.88	2.01	1.50	0	10.83	23.04
PS	0.91	37.44	57.28	4.37	91.14	4.09	0.09	0	0.31	36.29

* calculated by difference.
** calculated using a modified Dulong`s formula.

4. Conclusion

A sequential pyrolysis and catalytic reforming process in a continuous system has been successfully tested for the conversion of HDPE and PS into liquid and gaseous fuels using the HY-zeolite catalyst for the catalytic reforming of pyrolysis gas generated in a pyrolyzer. The effect of the reforming temperature and WHSV on the product yields, the liquid characteristics and the gaseous composition have been investigated for both HDPE and PS samples. There were significant influences of the reforming temperature and WHSV on the products yields for both HDPE and PS. Increasing the reforming temperature and decreasing WHSV have resulted in an increase of gaseous and solid products while the liquid product decreased. The maximum oil production for HDPE (70.0wt%) and PS (88.1wt%) were obtained at the pyrolysis temperature of 450 °C, the reforming temperature of 450 °C and WHSV of 4.

C_2, C_3 and C_{4+} gases (>75 mol %) and valuable aromatic and branched species in the light oil range (C_5-C_{15}) (>70wt %) were the main components of the gaseous and liquid products for HDPE, which were attributed to the molecule structure of HDPE, as well as the relatively moderate acidity and large hole size of the HY-zeolite catalyst. In case of PS, C_2 and C_3 gases (>65 mol %) were the main components of the gaseous product. However, the quantity of gaseous products was very low to serve as a fuel gas either for engine or for the SPCR process. In addition, the liquid products of PS consisted of mostly monocyclic aromatic hydrocarbon, such as styrene and benzene, as well as lower concentrations of potentially harmful polycyclic aromatic hydrocarbons. The high quality gaseous product can be used as a fuel either for driving gas engines or for dual-fuel diesel engines. It can also be used as a heating source for the pyrolysis reactor.

Acknowledgements

The authors would like to thank Dr. Tohru Kamo of AIST, Tsukuba, Japan for his valuable help and discussion.

References

Aguado, J., Serrano, D. P., Miguel, G. S., Castro, M. C., & Madrid, S. (2007). Feedstock recycling of polyethylene in a two-step thermo-catalytic reaction system. *Journal of Analytical and Applied Pyrolysis, 79*,415-423. http://dx.doi.org/10.1016/j.jaap.2006.11.008

Aguado, J., & Serrano, D. P. (2006). Catalytic Upgrading of Plastic Wastes. *Feedstock Recycling and Pyrolysis of Waste Plastics: Converting Waste Plastics into Diesel and Other Fuels.* 73-110. http://dx.doi.org/10.1002/0470021543.ch3

Aguado, J., & Serrano, D. P. (1999). *Feedstock recycling of plastic wastes* (Vol. 1). Royal society of chemistry.

Al-Khattaf, S. (2002). The influence of Y-zeolite unit cell size on the performance of FCC catalysts during gas oil catalytic cracking. *Applied Catalysis A: General, 231,* 293-306. http://dx.doi.org/10.1016/S0926-860X(02)00071-6

Audisio, G., & Bertini, F. (1990). Catalytic degradation of polymers: Part III—Degradation of polystyrene. *Polymer Degradation and Stability, 29,* 191-200. http://dx.doi.org/10.1016/0141-3910(90)90030-B

Bagri, R., & Williams, P. T. (2002). Catalytic pyrolysis of polyethylene. *Journal of Analytical and Applied Pyrolysis, 63,* 29-41. http://dx.doi.org/10.1016/S0165-2370(01)00139-5

Buekens, A. G., & Huang, H. (1998). Catalytic plastics cracking for recovery of gasoline-range hydrocarbons from municipal plastic wastes. *Resources, Conservation and Recycling, 23,* 163-181. http://dx.doi.org/10.1016/S0921-3449(98)00025-1

Castanoa, P., Elordia, G., Olazara, M., Aguayoa, A. T., Pawelecb, B., & Bilbaoa, J. (2011). Insights into the coke deposited on HZSM-5, Hβ and HY zeolites during the cracking of polyethylene. *Applied Catalysis B: Environmental,104,* 91-100. http://dx.doi.org/10.1016/j.apcatb.2011.02.024

Chumbhale, V. R., Kim, J. S., Lee, W. Y., Song, S. H., Lee, S. B., & Choi, M. J. (2005). Catalytic Degradation of Expandable Polystyrene Waste (EPSW) over HY and Modified HY Zeolites. *Journal of Industrial and Engineering Chemistry, 11*(2), 253-260.

Demirbas, A. (2010). *Fuels from Biomass,* Biorefineries. Springer London (pp.33-73).

Iribarren, D., Dufour, J., & Serrano, D. P. (2012). Preliminary assessment of plastic waste valorization via sequential pyrolysis and catalytic reforming. *Journal of Material Cycles and Waste Management, 14*(4), 301-307. http://dx.doi.org/10.1007/s10163-012-0069-6

Ishaq, M., Ahmad, I., Shakirullah, M., Khan, M.A., Rehman, H., & Bahader, A. (2006). Pyrolysis of some whole plastics and plastics-coal mixtures. *Energy Conversion and Management, 47*, 3216-3223. http://dx.doi.org/10.1016/j.enconman.2006.02.019

Joo, H. S., & Guin, J. A. (1997). Hydrocracking of a plastics pyrolysis gas oil to naphtha. *Energy and Fuels,11*, 586-592. http://dx.doi.org/10.1021/ef960151g

Kiran, N., Ekinci, E., & Snape, C. E. (2000). Recycling of plastic wastes via pyrolysis, *Resources, Conservation and Recycling, 29*, 273-283. http://dx.doi.org/10.1016/S0921-3449(00)00052-5

Lee, S. Y., Yoon, J. H., & Park, D. W. (2002). Catalytic degradation of mixture of polyethylene and polystyrene. *Journal of Industrial and Engineering Chemistry, 8*(2), 143-149.

Lin, H. T., Huang, M. S., Luo, J. W., Lin, L. H., Lee, C. M., & Ou, K. L. (2010). Hydrocarbon fuels produced by catalytic pyrolysis of hospital plastic wastes in a fluidizing cracking process. *Fuel Processing Technology, 91*, 1355-1363. http://dx.doi.org/10.1016/j.fuproc.2010.03.016

Liu, Y., Qian, J., & Wang, J. (2000). Pyrolysis of polystyrene waste in a fluidized-bed reactor to obtain styrene monomer and gasoline fraction, *Fuel Processing Technology, 63*, 45–55. http://dx.doi.org/10.1016/S0378-3820(99)00066-1

Luo, G. H., Suto, T., Yasu, S., & Kato, K. (2000). Catalytic degradation of high density polyethylene and polypropylene into liquid fuel in a powder-particle fluidized bed. *Polym Degrad Stabil,70*, 97-102. http://dx.doi.org/10.1016/S0141-3910(00)00095-1

Manos, G., Garforth, A., & Dwyer, J. (2000). Catalytic degradation of high-density polyethylene over different zeolitic structures. *Industrial & engineering chemistry research, 39*(5), 1198-1202. http://dx.doi.org/10.1021/ie990512q

Marcilla, A., Beltran, M. I., & Navarro, R. (2009). Thermal and catalytic pyrolysis of polyethylene over HZSM5 and HUSY zeolites in a batch reactor under dynamic conditions. *Applied Catalysis B: Environmental, 86*, 78-86. http://dx.doi.org/10.1016/j.apcatb.2008.07.026

Mikulec, J., & Vrbova, M. (2008). Catalytic and thermal cracking of selected polyolefins. *Clean Techn Environ Policy, 10*, 121-130. http://dx.doi.org/10.1007/s10098-007-0132-5

Miskolczi, N., Bartha, L., Deak, G., Jover, B., & Kallo, D. (2004). Thermal and thermo-catalytic degradation of high-density polyethylene waste. *Journal of Analytical and Applied Pyrolysis, 72*, 235-242. http://dx.doi.org/10.1016/j.jaap.2004.07.002

Murata, K., Brebu, M., & Sakata, Y. (2010). The effect of silica-alumina catalysts on degradation of polyolefins by a continuous flow reactor. *Journal of Analytical and Applied Pyrolysis, 89*, 30-38. http://dx.doi.org/10.1016/j.jaap.2010.05.002

Murata, K., Brebu, M., & Sakata, Y. (2009). Thermal degradation of polyethylene into fuel oil over silica–alumina by continuous flow reactor. *Journal of Analytical and Applied Pyrolysis, 86*, 354-359. http://dx.doi.org/10.1016/j.jaap.2009.08.009

Neves, I. C., Botelho, G., Machado, A. V., & Rebelo, P. (2007). Catalytic degradation of polyethylene: An evaluation of the effect of dealuminated Y zeolites using thermal analysis. *Materials Chemistry and Physics, 104*, 5-9. http://dx.doi.org/10.1016/j.matchemphys.2007.02.032

Neves, I. C., Botelho, G., Machado, A. V., & Rebelo, P. (2006). The effect of acidity behaviour of Y zeolites on the catalytic degradation of polyethylene. *European Polymer Journal, 42*, 1541-1547. http://dx.doi.org/10.1016/j.eurpolymj.2006.01.021

Onwudili, J. A., Insura, N., & Williams, P. T. (2009). Composition of products from the pyrolysis of polyethylene and polystyrene in a closed batch reactor: Effects of temperature and residence time. *Journal of Analytical and Applied Pyrolysis, 86*, 293-303. http://dx.doi.org/10.1016/j.jaap.2009.07.008

Park, Y., Namioka, T., Sakamoto, S., Min, T. J., Roh, S., & Yoshikawa, K. (2010). Optimum operating conditions for a two-stage gasification process fueled by polypropylene by means of continuous reactor over ruthenium catalyst. *Fuel Processing Technology, 91*, 951-957. http://dx.doi.org/10.1016/j.fuproc.2009.10.014

Pinto, F., Costa, P., Gulyurtlu, I., & Cabrita, I. (1999). Pyrolysis of plastic wastes. 1. Effect of plastic waste composition on product yield. *Journal of Analytical and Applied Pyrolysis, 51*, 39-55. http://dx.doi.org/10.1016/S0165-2370(99)00007-8

Puente, G., & Sedran, U. (1998). Recycling polystyrene into fuels by means of FCC: performance of various acidic catalysts. *Applied Catalysis B: Environmental, 19,* 305-311. http://dx.doi.org/10.1016/S0926-3373(98)00084-8

San Miguel, G., Serrano, D. P., & Aguado, J. (2009). Valorization of waste agricultural polyethylene film by sequential pyrolysis and catalytic reforming. *Industrial & Engineering Chemistry Research, 48*(18), 8697-8703. http://dx.doi.org/10.1021/ie900776w

Seo, Y. H., Lee, K. H., & Shin, D. Y. (2003). Investigation of catalytic degradation of high-density polyethylene by hydrocarbon group type analysis. *Journal of Analytical and Applied Pyrolysis, 70,* 383-398. http://dx.doi.org/10.1016/S0165-2370(02)00186-9

Serrano, D. P., Aguado, J., & Escola, J. M. (2000). Catalytic conversion of polystyrene over HMCM-41, HZSM-5 and amorphous SiO_2–Al_2O_3: comparison with thermal cracking. *Applied Catalysis B: Environmental, 25,* 181–189. http://dx.doi.org/10.1016/S0926-3373(99)00130-7

Uemichi, Y., Hattori, M., Itoh, T., Nakamura, J., & Sugioka, M. (1998). Deactivation behaviors of zeolites and silica-alumina catalysts in the degradation of polyethylene. *Ind Eng Chem Res, 37,* 867-872. http://dx.doi.org/10.1021/ie970605c

UNEP, *Converting waste plastics into resource: compendium of technologies,* Osaka, 2009.

Wang, J. L., & Wang, L. L. (2011). Catalytic pyrolysis of municipal plastic waste to fuel with nickel-loaded silica-alumina catalysts. *Energy Sources Part A: Recovery, utilization and environmental effects, 33,* 1940-1948.

Williams, P. T., & Bagri, R. (2004). Hydrocarbon gases and oils from the recycling of polystyrene waste by catalytic pyrolysis. *International Journal of Energy Research, 28*(1), 31-44. http://dx.doi.org/10.1002/er.949

Williams, P. T., & Slaney, E. (2007). Analysis of products from the pyrolysis and liquefaction of single plastics and waste plastic mixtures. *Resource Conservation Recycling, 51,* 754-769. http://dx.doi.org/10.1016/j.resconrec.2006.12.002

Williams, P. T., & Williams, E. A. (1999). Interaction of plastics in mixed plastics pyrolysis. *Energy and Fuels, 13,* 188-196. http://dx.doi.org/10.1021/ef980163x

Zhang, Z., Hirose, T., Nishio, S., Morioka, Y., Azuma, N., Ueno, A., & Okada, M. (1995). Chemical recycling of waste polystyrene into styrene over solid acids and bases. *Industrial & engineering chemistry research, 34*(12), 4514-4519. http://dx.doi.org/10.1021/ie00039a044

Zhou, L., Luo, T., & Huang, Q. (2009). Co-pyrolysis characteristics and kinetics of coal and plastic blends. *Energy Conversion and Management, 50,* 705-710. http://dx.doi.org/10.1016/j.enconman.2008.10.007

4

Environmental Risk Due to Heavy Metal Contamination Caused by Old Copper Mining Activity at Ľubietová Deposit, Slovakia

Peter Andráš[1,2], Ingrid Turisová[1], Eva Lacková[3], Sherif Kharbish[4], Jozef Krnáč[1] & Lenka Čmielová[3]

[1] Faculty of Natural Sciences, Matej Bel University, Banská Bystrica, Slovakia

[2] Geological Institute of Slovak Academy of Sciences, Banská Bystrica, Slovakia

[3] VŠB-Technical University of Ostrava, Ostrava, Czech Republic

[4] Geology Department, Faculty of Science, Suez University, Suez Governate, El Salam City, Egypt

Correspondence: Peter Andráš, Geological Institute of Slovak Academy of Sciences, Banská Bystrica, Slovakia. E-mail: andras@savbb.sk

Abstract

The more than 200 years old dump-fields at closed Cu (Ag) deposit Ľubietová are situated near the village settlement. Heavy metal space distribution is controlled by geochemical behaviour of the elements, depend on their content, solubility, migration and sorption ability. The major sources of metals to the country components (soil, technogenous sediments, groundwater, surface water, plants…) may be classified according to expected solubility of primary minerals. The content of heavy metals in sediments and soils at the studied dump-field shows irregular distribution. Also the heavy metal contamination of the surface water and groundwater was studied both in the rainy as well as during the dry periods. The speciation of As and Sb proved in the water presence both of As^{3+} and Sb^{3+} as well as the less toxic As^{5+} and Sb^{5+} species. In the soil and sediments prevail As^{5+} and Sb^{5+} species while in the water is often dominant the As^{3+} and Sb^{3+} form. The article also presents some results of the plant tissue degradation study under heavy metal contaminated conditions at dump-fields. The dump sediments and the primitive soil formed locally on the surface of the technogenous sediments show only limited acidification potential. The Fe^0-barrier installation at bottom of the down part of the valley seems to be a good solution for the groundwater decontamination.

Keywords: acidification, heavy metal contamination, dump-field, impact on flora, remediation

1. Introduction

The Ľubietová deposit was exploited since the time of the Bronze Age and in the 16th and 17th centuries it was one of the most important and most extensively exploited Cu-mines of Europe. The Cu-ore was in the 18. century exported to more than 50 countries (Bergfest, 1951). The Cu mineralisation with Ag admixture is developed within 4–5 km long and 1.5 km wide range of N-S direction. It is situated in a crystalline complex which consists of arcose greywackes, arcose schists and various conglomerates. The ore mineralisation was genetically connected with the basic, intermediate and acid Permian volcanism. It is characterised by a rather simple paragenesis represented by quartz, siderite, chalcopyrite, Ag-tetrahedrite, arsenopyrite, pyrite and rare galena. The deposit is famous also by the numerous secondary minerals as libethenite, annabergite, langite, azurite, brochantite, erithrine, evansite, euchroite, cuprite, malachite, olivenite, pseudomalachite, etc. (Koděra et al., 1990; Ebner et al., 2004). There are three great ore-fields at Ľubietová: Podlipa, Svätodušná and Kolba with admixture of Co/Ni-mineralisation. The Cu content in the ore ranged from 4–10% and the Ag content was about 70 g·t⁻¹ (Bergfest, 1951).

Figure 1. Dump-field podlipa at abandoned Cu-deposit Ľubietová

The extent of the dump-field Podlipa (Figure 1) is about 2 km^2. This ore-field was exploited by 18 adits. The Cu-ore mineralisation is situated mainly in the terrigene crystalline complex of Permian age which consists of arcose greywackes, arcose schists and various conglomerates. The most important tectonic structures are of NE-SW direction and the ore veins have E-W and N-S direction. They are 30–40 m thick.

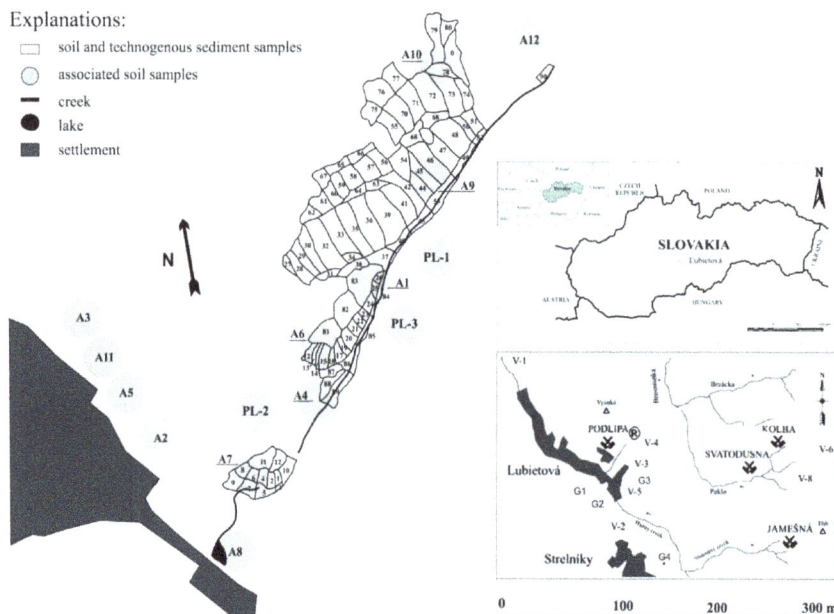

Figure 2. The Podlipa dump-field: localisation of the soil (samples A-1 to A-12 and 1–80) and water (samples V, CD and LH) sampling

The important mining stopped in the second half of the 19th century although the last little gallery near Haliar locality was exploited by Ernest Schtróbl only in April 1915.

2. Material and Methods

The technogenous sediment and soil samples (of about 10 kg weight; samples 1–80 and A-1 to A-12 from the Podlipa dump-field and soils from 15–20 cm depth (the sampling step was about 25 m^2) were collected in order to characterize components of landscape contamination. Next three samples PL-1 to PL-3 are samples of soil horizons (A, B, C). To each surface water sample (stream water, drainage water-samples V-1 to V-6) and groundwater samples (G-1 to G-4; Figure 2) 10 ml of HCl was added to conserve them.

The reference site was located above the ore-field (Figure 2, sample A-12), outside of geochemical anomalies of heavy metals and represent graywakes of Permian age similar to material at the dump-field. Plant samples were collected both from reference area and from the dump-area.

The samples of technogenous sediments and soil from the dumps were dried and 0.25 g of sample was heated in HNO_3-$HClO_4$-HF to fuming and taken to dryness. The residue was dissolved in HCl. Solutions were analysed by ICP-MS analysis at the ACME Analytical Laboratories (Vancouver, Canada). The minerals in the clay fractions were determined by X-ray diffraction and the clay mineral samples were ICP-MS analysed, then macerated 14 days in natural drainage water of the studied locality containing heavy metals and analysed again. The pH of the sediments was determined from suspension both with distilled water and 1M KCl after 3 hours of maceration.

The sulphur, total carbon, organic carbon and inorganic carbon content in the sediments was IR analysed using furnace Ströhlein C-MAT 4000 at the laboratories of Geological Institute of Slovak Academy of Sciences. A static test of the total acid potential was realized according to Morin and Hutt (1997) and Sobek et al. (1978). The water samples were analysed using AAS in the National Water Reference Laboratory for Slovakia at the Water Research Institute in Bratislava. The speciation of As was performed on the basis of different reaction rate of As^{3+} and As^{5+} depending on pH and Eh. The experimental study on Cu precipitation on the surface of iron particles (testing of the Fe^0-barrier) mixed with dolomite was realized at the laboratory of the Comenius University in Bratislava by Dr. Bronislava Lalinská according to Bartzas (2006).

3. Results

3.1 Heavy Metal Contamination of Technogenous Sediments and Soil

The dump-field technogenous sediments are influenced by heavy metals (Table 1) from the hydrothermal ore mineralisation. The main contaminants are: Cu (up to 20 360 ppm), Fe (up to 2.58%), As (up to 457 ppm), Sb (up to 79.3 ppm) and Zn (up to 80 ppm) are accompanied also by U (up to 10 ppm) and Th (up to 35 ppm) from the Permian volcano-sedimentary metamorphosed rocks.

Table 1. ICP-MS analyses of technogenous sediments and soils from the dump-field

Sample	Cu	Pb	Zn	Ni	Co	As	Sb	Bi	U
					ppm				
A-1	2829	28.1	14	36.8	10.4	162	61.6	2.8	1.3
A-1c	1693	63.8	18	36.0	11.3	258	60.1	4.5	1.4
A-1c*	2345	229.1	95	71.8	18.3	628	153.2	14.6	3.3
A-2	198.8	13.0	21	9.8	5.9	10	7.1	0.2	1.4
A-2c	574.3	22.4	36	12.2	10.3	19	9.2	1.4	1.1
A-2c*	472.4	27.9	62	17.0	6.4	15	12.6	1.5	1.1
A-3	827.5	16.0	20	32.1	14.0	71	22.4	8.5	1.7
A-3c	624.2	23.1	25	28.3	17.0	110	24.0	7.2	1.8
A-3c*	857.4	37.4	47	30.4	11.0	105	28.0	12.1	1.9
A-4	4471	9.6	23	55.0	50.0	169	59.5	23.7	1.6
A-4c	3324	14.9	16	42.4	58.3	237	79.3	39.2	1.7
A-4c*	3112	37.8	27	64.4	32.1	300	129.8	90.9	2.2
A-5	3150	16.9	19	34.0	24.4	60	17.2	1.7	1.0
A-5c	3001	14.8	18	34.1	30.4	64	16.3	2.1	1.2
A-5c*	2078	21.9	45	55.4	29.6	105	30.3	3.2	1.4
A-6	4797	15.6	13	51.6	41.8	134	49.8	25.4	1.4
A-6c	2503	24.6	14	45.1	40.9	224	56.2	24.4	1.6
A-6c*	2918	72.3	65	61.7	32.0	305	92.3	51.7	2.2
A-7	755.8	16.8	26	10.4	10.2	16	11.5	0.9	1.1
A-7c	855.1	20.2	33	10.1	12.0	17	7.1	1.2	1.1
A-7c*	2026	73.7	176	26.0	15.5	33	17.4	3.6	2.3
A-8	716.0	6.5	7	58.0	89.9	61	17.9	0.5	2.6
A-8c	835.5	6.3	14	66.5	69.7	52	20.2	0.7	2.5
A-8c*	836.7	4.2	4	62.5	104.5	46	18.9	0.8	2.1
A-9	5903	29.5	24	39.8	36.0	244	37.0	15.1	2.7
A-10	7699	30.2	19	52.2	48.0	457	62.7	25.1	4.0
A-11	1563	24.8	37	19.0	8.7	16	14.9	4.8	1.7
A-12	113.1	39.4	29	8.9	8.6	16	5.6	0.7	1.4
A-17	14 440	8.4	59	51.7	73.4	289	43.2	7.2	2.3
A-17c	20 360	49.0	80	43.0	70.0	260	40.0	6.0	2.0

Explanations: A-1 to A-11 – technogenous sediments and soils from the dump-field, A-12 reference area, A-17 hydrogoethite, A-1c to A-17c clay fraction, A-1* to A-17* - clay fraction after 14 days maceration in drainage water, containing heavy metals.

The weathering processes of reactive minerals in surrounding and mainly acid rocks mobilise heavy metals and toxic elements (e.g. Cu, As) from the primary minerals (Figure 4), form secondary minerals (mainly Cu-oxides and arsenates – Figure 5, Fe hydroxides – Figure 6 and carbonates – Figure 7) and contaminate the landscape components.

The main Cu concentration was found to be near the Najvyššia Johan gallery collar, inclusive of the slope beneath the gallery. High Cu concentration is at the bottom of the valley (Figure 3a). The most important source of As and Fe are the galleries Najvyššia Johan, Horná Johan and Stredná Johan (Figures 3b, 3c). The highest Pb concentration was found in front of the Zollweiner Maria Empfängnis and Jakob galleries collars (Figure 3d). Th contamination is derived from Francisci and Bartolomej galleries (Figure 3e) and important Sb contamination was detected mainly at the dumps at Zollweiner Maria Empfängnis gallery (Figure 3f).

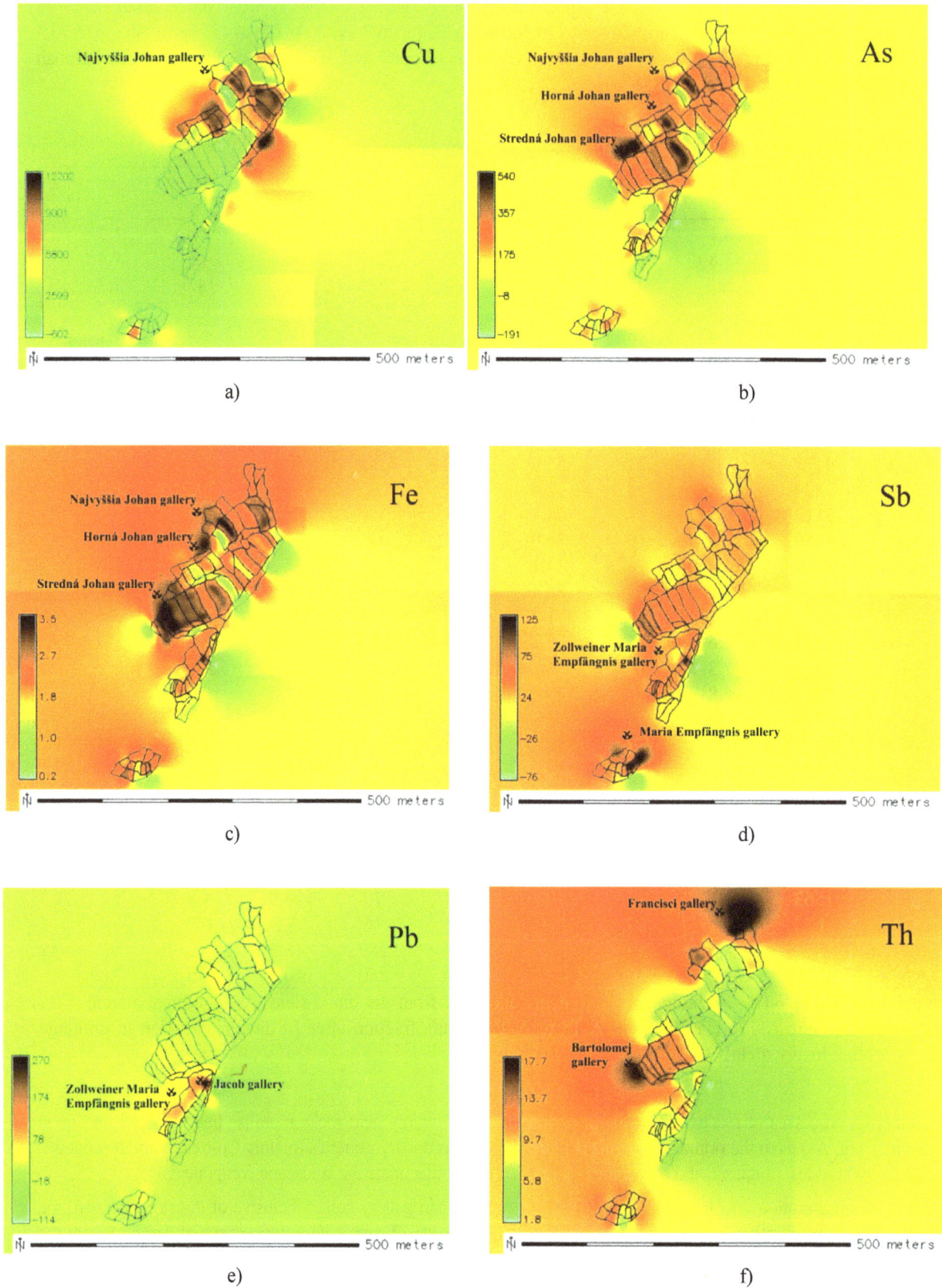

Figure 3. Distribution of heavy metals at the Podlipa dump-field: Cu, As, Fe, Pb, Th, Sb-distribution (numbers show the individual heavy metal content in ppm)

In mould horizon A are accumulated Ca, P, Pb, Zn, Cd, Cr and Th. In soil horizont B folowing elements: Cu, Ag, Ni and As are accumulated in clay minerals and sesquioxides. The C horizon is enriched in Ba (Table 2). The other analysed elements (e.g. Fe, Mg, Ti, Al, Na, K, Mo, Mn, U) show no unambiguous trend to be accumulated in some soil horizon. The Sb, Bi and Co concentrations in individual soil horizons show no unambiguous trends.

Table 2. ICP-MS analyses of individual soil horizons

Sample	Horizon	Fe	Ca	P	Mg	Ti	Al	Na	K
						%			
PL-1	A	2.48	0.40	0.093	0.62	0.155	5.98	0.328	3.02
	B	2.01	0.27	0.090	0.59	0.125	6.01	0.269	3.37
	C	2.05	0.17	0.067	0.50	0.146	6.41	0.344	3.74
PL-2	A	3.02	0.33	0.097	0.42	0.099	5.81	0.321	3.05
	B	2.49	0.22	0.058	0.45	0.084	6.06	0.496	3.43
	C	6.87	0.12	0.053	0.26	0.093	5.73	0.122	3.01
PL-3	A	1.32	0.49	0.092	0.56	0.083	4.45	0.052	2.24
	B	1.65	0.05	0.076	0.44	0.097	5.58	0.043	3.03
	C	2.01	0.11	0.077	0.66	0.154	4.73	0.022	4.01

Sample	Horizon	Cu	Pb	Zn	Ag	Ni	Cd	As	Sb
						ppm			
PL-1	A	5864	32.6	136	7.6	38	0.7	1010	2449
	B	9425	33.3	130	8.5	40.9	0.4	1081	2356
	C	3633	10.4	89	4.7	36.7	0.2	497	1159
PL-2	A	4462	33.9	133	5.3	27.5	1.1	553	1216
	B	8230	17.2	134	6.8	33.6	0.5	884	1539
	C	6059	8.1	142	10	33	0.3	565	1704
PL-3	A	3366	66	77	1.1	30.8	1.1	77	24
	B	8085	15.1	11	2.2	36.3	0.1	231	23
	C	5115	15.1	16	1.2	34.3	0.2	158	37

Sample	Horizon	Mo	Mn	Co	Bi	Cr	Ba	U	Th
						ppm			
PL-1	A	0.7	1033	46.5	122	31	607	5.4	9.2
	B	1.2	953	43.8	112.1	26	694	6.1	6.9
	C	0.7	765	26.5	42.9	19	705	3.3	7.3
PL-2	A	0.9	1123	32.8	49.4	22	620	3.2	7.5
	B	0.4	1171	49.2	70.3	13	841	3.2	6.3
	C	0.7	2220	20.1	37.4	20	1099	3.5	7.3
PL-3	A	3.3	352	35.2	3.3	18	218	1.4	4.9
	B	3.3	253	45.1	8.1	11	283	2.9	6.7
	C	0.3	351	48.8	8.6	13	303	3.3	7.7

The distribution of individual metals at the studied dump-field is uneven. It depends on the mineralogical composition of ores, on the original concentration of the mentioned metals in the technogenic sediments of the spoil dumps, and also on their migration abilities and sorption properties (Figure 3). At Podlipa were described only few correlations between metal couples (Ni/As, Fe/Ni, Pb/Cd, Co/Cu., Zn/Cd, Zn/Pb, Cd/Pb ant Th/U). This situation substantially differs from the dump-field Reiner, where we can distinguish high correalition ratio between four times more element couples as at Podlipa (Andráš et al., 2009). The main reasons of this separate geochemical behaviour of the same elements at the two neighbour localities is the different rock-surrounding (homogenous rock-surrounding at Reiner locality built by greywakes vs. grauwakes - arcose schists rock composition at Podlipa, which form different quantity of natural sorbents as clay minerals; Andráš et al. (2009). The distribution of elements can be controlled also by formation of various metastable, secondary phases, which depend on pH and Eh, eventually on ionic radius of elements.

Figure 4. Grain of native copper

Figure 5. Euchoite ($Cu_2(AsO_4)(OH) \cdot 3H_2O$) aggregate

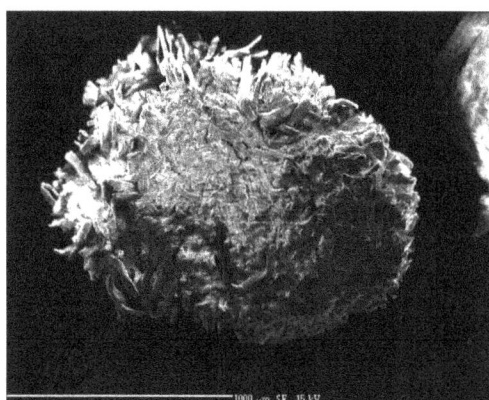

Figure 6. Lepidocrocite (γ-FeO(OH)) aggregat

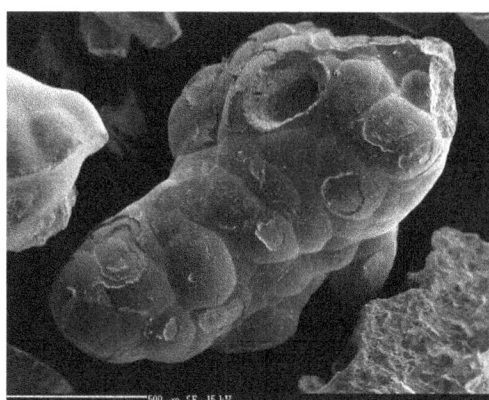

Figure 7. Malachite ($Cu_2CO_3(OH)_2$) aggregate

3.2 Heavy Metals in Water

The surface water in the creek draining the valley bottom along the dump-field is gradually contaminated by heavy metals leached from the technogenous sediments of the mining dumps. This surface water contains high Cu (up to 2060 $\mu g \cdot L^{-1}$), Fe (up to 584 $\mu g \cdot L^{-1}$), Zn (up to 35 $\mu g \cdot L^{-1}$) and sometimes also Co (up to 10 $\mu g \cdot L^{-1}$) and Pb (up to 5 $\mu g \cdot L^{-1}$) concentrations. The highest As concentration is 6.11 $\mu g \cdot L^{-1}$ (Table 3).

Table 3. Atom absorption spectrometric analyses of surface water

Sample	pH	Eh	Mn	Zn	Cd	Co	Cu	Fe	Ni	Pb	Sb	As
		(mV)					$\mu g.l^{-1}$					
V-1a	6.5	-6	<1	<10	0.04	1.1	2.2	26	4.1	4.2	0.74	<1.0
V-1b	7.5	-58	<1	<10	0.05	2.2	2.7	73	5.9	4.3	<1.00	<1.0
V-1c	6.5	-8	11	<10	<0.05	<1.0	5.1	94	1.2	<1.0	1.03	<1.0
V-2b	6.7	-14	<1	<10	0.13	<1.0	42.1	584	2.1	3	<1.00	1.69
V-2c	6.9	-21	<1	<10	0.09	<1.0	38.2	580	1.6	2.9	<1.00	1.54
V-3a	6.7	-12	<1	30	0.04	7	1 810	86	3.2	2.2	1.12	<1.00
V-3b	6.1	14	<1	40	0.05	9.6	2 060	101	4.9	2.8	1.88	3.41
V-3c	6.5	0	21	<10	<.05	7.6	1 980	45	8.5	2.8	2.35	1.14
V-4a	6.7	-14	<1	<10	0.06	3.1	22.2	263	2.1	4.2	1.72	<1.0
V-4b	6.2	14	<1	<10	0.06	8.1	1 850	274	5.6	3.6	1.57	1.21
V-5a	6.2	-11	<1	<10	0.06	5.5	6	170	6	4.8	1.66	2.79
V5b	6.3	-8	7	20	0.08	8.3	7.9	210	7.1	5.1	2.21	3.21
V-5d	6.2	-7	4	30	0.07	6.6	8.1	160	8.1	1	2	1.08
V-6a	7.6	-62	<1	30	0.07	1.9	30.4	270	4.3	3.2	2	6.02
V-6b	7.1	-62	<1	32	0.07	2.2	34.8	263	5	3.4	2.01	6.11

Explanations: Samples marked by index "a" - rainy period (June 14th, 2006), samples marked by index "b" - dry period (February 25th, 2007), samples marked by index "c" - rainy period (March 31st 2008), samples marked by index "d" - dry period (May 27th, 2008).

The heavy metals content in the water is in most cases higher during dry periods than during rainy periods. On the other hand, the As content both in the surface (and drainage) water as well as in the groundwater is not high (6.11 $\mu g \cdot L^{-1}$; Table 3). The presence of *Acidithiobacteria* species or of sulphate reducing bacteria was not proved. The pH both of the surface and of groundwater is close to neutral values (pH 6.4–7.6).

The most contaminated is the mineral water from the spring Linhart (sample G-4; Figure 2). Its total radioactivity is 6,498 $Bq \cdot L^{-1}$ and the Fe (380 $\mu g \cdot L^{-1}$), Cu (181 $\mu g \cdot L^{-1}$), Pb (1 $\mu g \cdot L^{-1}$) and Cd (82.0 $\mu g \cdot L^{-1}$) contents (Table 4) exceed the Slovak decrees No. 296/2005, No 354/2006 Coll.

Table 4. Atom absorption spectrometric analyses of groundwater (samples G-1,G-2 and G-3) and mineral water (sampleG-4)

Sample	pH	Eh (mV)	Fe	Ni	Mn	Zn	Cu	Cd	Pb	Bi	As	Sb
							$\mu g.l^{-1}$					
G-1a	6.55	-4	11	1.1	<5	<10	22	<0.05	<1.1	1.36	<1	<1.0
G-1b	6.63	-10	17	1.2	<5	<10	1.3	<0.05	1.9	1.55	<1	<1.0
G-2a	6.72	-14	366	1.3	18	<10	3	<0.05	1.3	<1.00	<1	<1.0
G-2b	6.23	-16	210	1.5	8	<10	2.2	<0.05	<1.0	<1.00	<1	<1.0
G-3a	6.85	-21	146	<1.0	15	61	14	0.13	3.4	<1.00	5	1.42
G-3b	6.55	-23	120	<1.0	17	350	5.9	0.1	3.3	<1.00	1.52	1.21
G-4a	6.4	4	380	5	20	<10	30	0.5	<1.0	<1.00	1.98	<1.0
G-4b	6.48	-2	2 260	<1.0	55	<10	181	82	<1.0	<1.00	2.52	<1.0

Explanations: a - sampled on March 31st 2008 during the rainy period; b – sampled on May 27th 2008 during the dry period.

Table 5. Paste and rinse pH (H₂O and 1 M KCl), sulphur and carbon contents in samples of technogenous sediments and soils

Sample	pH_{H2O}	Eh_{H2O}	pH_{KCl}	Eh_{KCl}	$S_{tot.}$	S_{SO4}	S_s	$C_{tot.}$	$C_{org.}$	$C_{inorg.}$	CO_2	$CaCO_3$	TAP $(kg.t^{-1})$
A-1	5.14	77	4.61	109	0.25	0.1	0.15	0.74	0.2	0.54	1.97	4.48	7.813
A-2	5.89	34	5.4	63	0.02	0.01	0.01	0.86	0.38	0.48	1.75	3.99	0.625
A-3	4.87	94	4.21	131	0.1	0.03	0.07	0.62	0.34	0.28	1.02	2.32	3.125
A-4	5.46	59	5.33	66	0.33	0.13	0.01	0.34	0.26	0.08	0.29	0.66	10.313
A-5	5.77	42	5.37	64	0.05	0.01	0.05	0.78	0.35	0.43	1.57	3.57	1.563
A-6	5.17	74	5.06	83	0.42	0.15	0.27	0.4	0.27	0.13	0.47	1.08	13.125
A-7	7.93	-84	7.34	-58	0.03	0.02	0.01	1.63	0.1	1.53	5.61	12.71	0.938
A-8	5.42	36	5.22	42	0.01	0.01	0.01	0.45	0.13	0.32	1.17	2.66	0.313
A-9	5.03	83	5.01	85	0.03	0.03	0.01	0.4	0.37	tr.	tr.	tr.	0.938
A-10	5.25	71	5.14	78	0.04	0.02	0.02	0.48	0.46	tr.	tr.	tr.	1.25
A-11	6.11	22	5.95	30	0.11	0.04	0.07	4.31	4.18	0.13	0.47	1.08	3.438
A-12	4.21	133	3.47	173	0.02	0.01	0.02	4.05	4.03	tr.	tr.	tr.	0.625

In spite of the limited kinetics of the cementation process, the electron microprobe study proved that the cementation causes on the surface of iron particles gradual displacement of Fe^{2+} ions and precipitation of Cu^{2+} ions, both in form of Cu-oxides, Cu carbonates and in form of native copper. The cementation copper is of a high fineness (it contain up to 96.07 wt. %).

The pH-Eh stability diagram for Cu-Fe-S-H₂O system (Fairthorne et al., 1997) show that formation of insoluble Fe oxides/hydroxides is possible both at neutral or alcaline pH and at high Eh values also in acid conditions (Figure 8). According to Fairthorne et al. (1997) the decrease in the amount of metal ions present in solution when the pH is increased and when nitrogen is replaced with oxygen is due to the formation of iron and copper oxides/hydroxides at the mineral (chalcopyrite) surface creating a physical barrier for further metal dissolution. The Eh value measured in this study at pH 5 is very close to the separation line Fe^{2+}/Fe_2O_3 when the mineral is conditioned in nitrogen but it moves away from this line in oxygen conditioning. The formation of insoluble cupric oxide/hydroxide is less thermo-dynamically favourable and only occurs at pH values larger than 6.

The Cu in soil and in technogenous sediments is present mostly in Cu^{2+} form, less as a Cu^{3+} (Cu_2O_3 - stability field) and only in 2 cases in the elementary form (Cu^0). Cu from the surface water occupies several stability fields: Cu_2O_3, Cu^0, Cu_2S, $CuFeS_2$ as well as CuS fields. Similar situation is in the case of the groundwater with exception of the Cu_2O_3 field (Figure 8). It means that the Cu in surface water is present in Cu^0, Cu^{2+} and Cu^{3+} forms while in groundwater only in Cu^0 and Cu^{2+} forms.

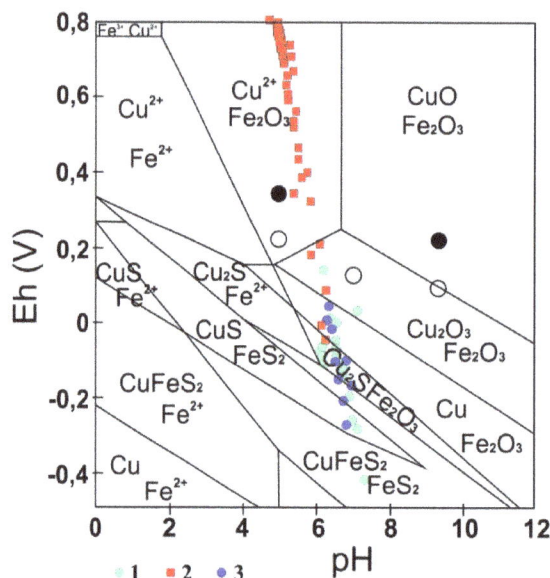

Figure 8. pH- Eh stability diagram for Cu-Fe-S-H_2O system (Fairthorne et al., 1997); (o) - the separation line Fe^{2+}/Fe_2O_3 in nitrogen conditions and ($\bullet$) - in oxygen conditions

The Fe in soil and in technogenous sediments is in the Eh-pH stability diagram for Cu-Fe-S-H_2O system (Fairthorne et al., 1997) situated only in the Fe_2O_3 stability field (Figure 8) or $Fe(OH)_3$ stability field (Figure 9), in consequence of this finding it is possible presume that the Fe is present in Fe^{3+} form. In groundwater and in surface water was described mostly Fe^{3+} but less often also Fe^{2+} form (Figure 8).

The speciation of As (Figure 9) and Sb (Figure 10) indicate that there are present both As^{3+}, Sb^{3+} and the less toxic As^{5+}, Sb^{5+} species. In the sediments prevail As^{5+} and Sb^{5+} species.

Migration of As and Sb in water may be realised in form of acidic or basic oxyions $H_2AsO_4^-$, $HAsO_4^{2-}$ and $HAsO_2^0$ or SbO_3^- and SbO_2^- under mildly oxidising conditions (Greenwood & Earnshaw, 1990; Manning & Goldberg, 2011). Under reducing, near-neutral to more alkaline conditions, transport of a sulphide complex such as $Sb_2S_4^{2-}$ is possible. By far the bulk of the stability field of water is covered by solid Sb-species (Sb_2S_3, $Sb(OH)_3$, Sb_2O_4, Sb_2O_5), thus suggesting that Sb transport must take place at moderately low Eh values.

At Ľubietová are in the water dominant the As^{3+} and Sb^{3+} forms (Figures 9, 10). As^{3+} is much more mobile than As^{5+} (Greenwood & Earnshaw, 1990; Manning & Goldberg, 2011) in weathering zone. The high As content in the water is controlled by tetrahedrite (and arsenopyrite) decomposition and by As sorption on amorphous Fe oxides and oxyhydroxides, on clay minerals and hydrogoethite (Andráš et al., 2007). Antimony is under supergenous conditions mobile and it has very similar behavior as arsenic Vink (1966). Determination of antimony speciation enables pH-Eh stability diagrams (Vink, 1966). Part of the anthropogenic sediment samples and soils are in the stability field of SbO_3^- water solution, the second part in senarmontite stability field (Figure 10).

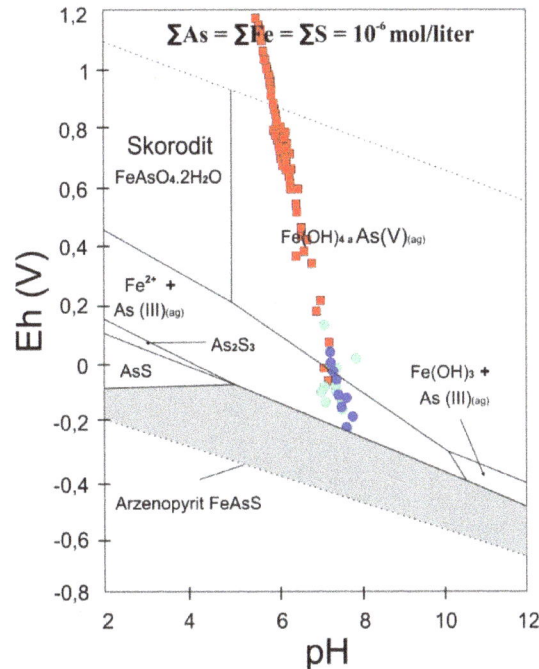

Figure 9. pH-Eh stability diagram of Fe-As-H_2O-S system according to Ryu (2002)

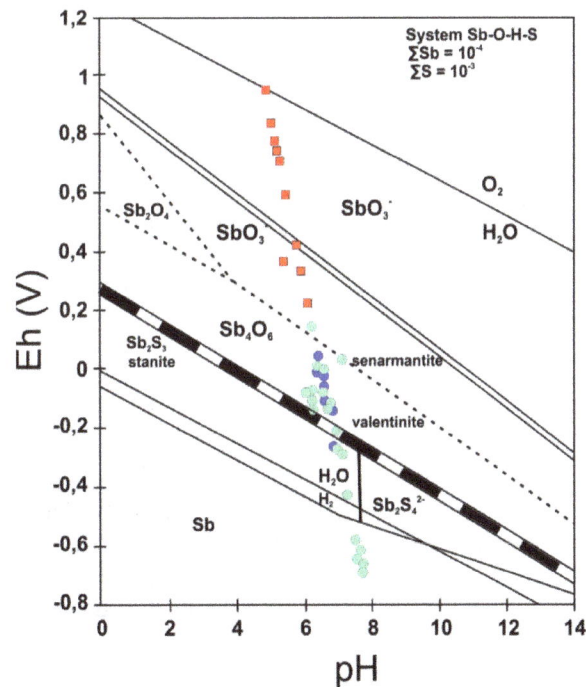

Figure 10. pH-Eh stability diagram of Sb-H_2O-S system according to according to Vink (1966)

In groundwater, antimony is mainly present in the form Sb_4O_6. Only in one single sample we can find antimony in the form of elementary Sb^0 and Eh as well as pH values of one sample are at diagram on the border between water solution and stibnite – Sb_2S_3 (Figure 12). Also in surface water the dominant form of antimony is Sb_4O_6. Forms SbO_3^- and elementary Sb^0 are very rare (Figure 12).

3.3 Natural Sorbents

Transport of heavy metals in form of mobile nanoparticles is influenced by natural sorbents. As the most important natural sorbents at the studied locality were described clay minerals (X-ray diffraction analyse proved presence of illite and muscovite mixture, caolinite but also minerals of smectite group and chlorite group) as well as hydrogoethite (Figure 11).

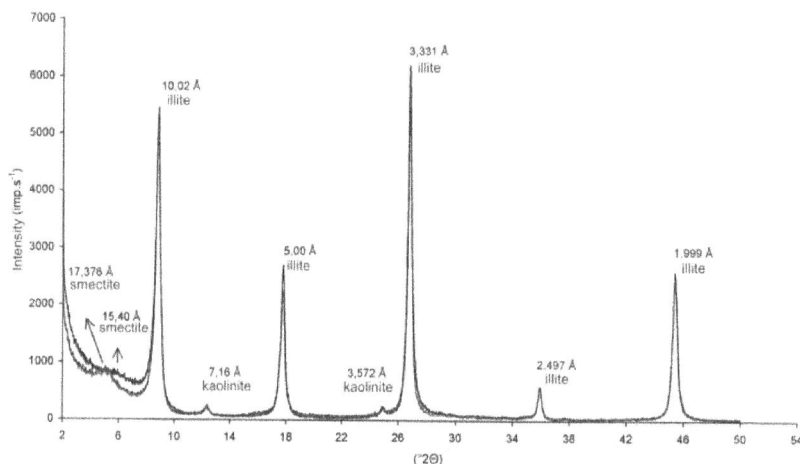

Figure 11. Rtg-diffraction diagram of clay mineral fraction from sample A-6

The best sorbent is hydrogoethite (Table 1, sample A-17). In most cases also the present clay minerals show good sorption capacity. These minerals generate in the dump-field material effective geochemical barrier which effect precipitation of metals and their fixation on the hydrogoethite and clay minerals surface.

Preferential sorption of Cr and Th on surface of clay minerals in comparison with hydroghoethite was described. On hydrogoethite surface are preferentially fixed Cu, Zn ($\pm$ Fe, Cd, Co).

Laboratory testing showed that most metals were sorbed on the clay minerals surface during 14 days maceration of the clay fraction in the drainage water percolating the dump-field material. This result indicates that the clay fraction still dispose by free sorption capacity.

Substantial differences in sorption capacity among various clay mineral mixtures were not certified, probably because of the matrix of all clay fraction consists predominantly of illite and muscovite. The variable quota of smectite, which has according to Andráš et al. (2007) higher sorption capacity as illite or muscovite, is not enough important to show substantially higher sorption efficiency.

3.4 The Total Acid Potential of the Dump-Field and the Posibility of Remediation

If distilled water is used in the measurement of paste or rinse pH of sediments or soils, its pH is usually around 5.3. pH values less than 5.0 indicates that the sample contains net acidity at the time of analyse (Sobek et al., 1978).

Values of paste pH between 5.0 and 10.0 can be considered near neutral at the time of the analysis. From the viewpoint of this study, only two samples (A-3 and A-12) account acid values (Table 5). It is surprising that one of these samples was taken from the reference area. The probable reason of such a behavior is the lack of carbonates (Table 5). The map of the soil and sediment acidity at the Podlipa dump-field is presented at Figure 12.

The measurement of the pH paste in the samples using solution of 1M KCl gives similar values. It means that only several few samples show markedly acid reaction (Table 5).

The very low total carbon ($C_{tot.}$) content (Table 5) reflects the lack of carbonates. The Figure 13 shows the $C_{tot.}$ distribution and enable comparison with the map of the acidity (Figure 12).

The total acid potential (TAP) was calculated according to Sobek et al. (1978):

$$TAP = \left(\% S_{tot.}\right) \times 31.25 \qquad (1)$$

where TAP is provided in any of three equivalent units: kg $CaCO_3$, equivalent/metric tone (t) of sample, t $CaCO_3$ equivalent/1000 t of sample, or parts per thousand (ppt) $CaCO_3$ equivalent.

The TAP values from the dump-field Podlipa range between 0.625 in samples A-12 (reference area) and A-2 to 13.125 in sample A-6 (Table 5).

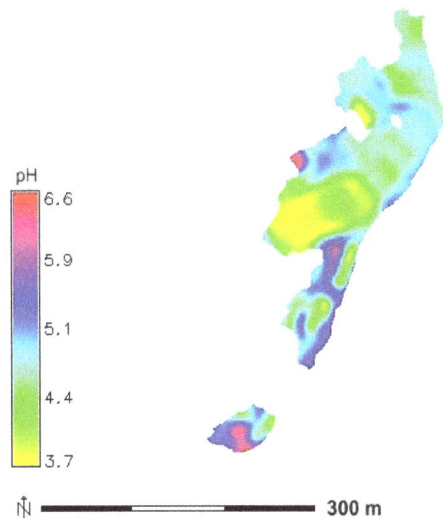

Figure 12. Map of the sediments acidity at the Podlipa dump-field

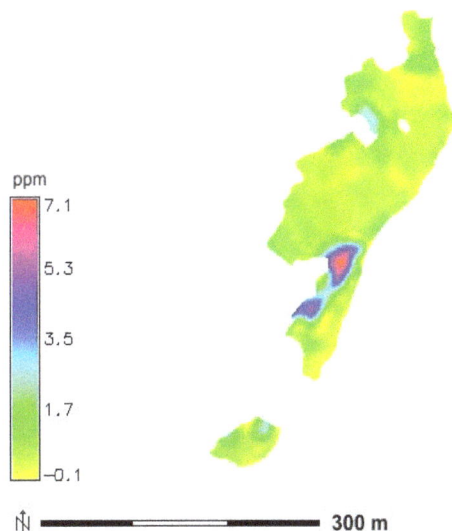

Figure 13. Map of the Ctot, distribution in the sediments at the Podlipa dump-field

If we compare the highest TAP value for sample A-6 with the data about the sulphur content, we can demonstrate that both the highest sulphide and sulphate sulphur contents were described from this sample, which represents the sedimentary material from Zollweiner Maria Empfängnis adit. The highest TAP value is also the consequence of the relatively low carbon content.

The ability of drainage water to precipitate cementation copper on iron surfaces makes the realization of an Fe^0-barrier for elimination of heavy metals (Cu, As, Cd, Zn and others) from the contaminated water a possibility. Mixture of Fe chips with dolomite also allowed to eliminate the released Fe.

3.5 Changes in Plant Tissues

We studied in 2009 and 2010 the influence of the specific ecological conditions at dump-fields with high heavy metal content on selected plant species which represent tolerant ecotypes. The relatively common *Pinus sylvestris* show various defects of their tissues: exfoliation of summer tracheide cell-wall layers, formation of the traumatic resin channels in closeness to the calluses, absence of the cell-wall coarsening or presence of funguses hyphaes in tracheae (Figure 14).

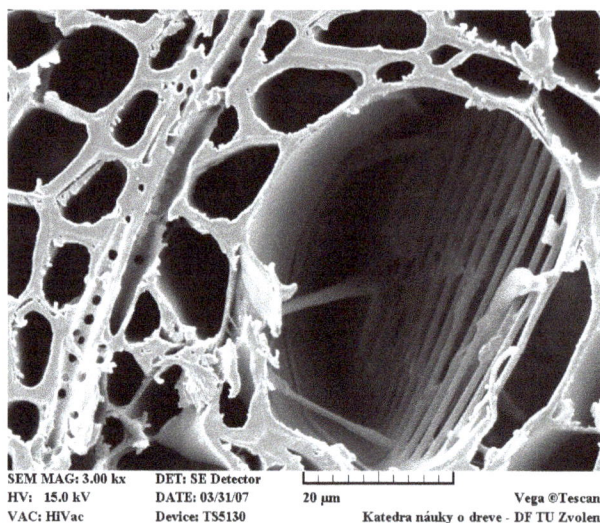

Figure 14. Funguses hyphaes in tracheae of Pinus sylvestris

Important chlorophyll insult causing even necrosis (mainly of leaves) was described in case of *Acetosella vulgaris*-plant which is considered to be so called „exclude-plants which use exclusion mechanisms by which uptake and/or root-to-shoot transport of heavy metals are restricted (Adams et al., 2011) or associated metal-tolerant species, which are moderately tolerant of heavy metals in soil, but not dependent on their presence (Baker et al., 2010). At 150 m long sectors from base of the dumps to their top we found 19 to 37 individuals of this species and at each plant were described on leaves violet to necrotic spots (Figure 15). Plants damage at individual sectors varied in range 88–100%.

Figure 15. Violet to necrotic spots on leaves of Acetosella vulgaris. a) initial damage; b) complete change of the original colour

Typical nanistic-pygmy forms, drying of branches, reduced organs (leaves, fruits), yellowing leaves, short annual shoot and dense age-bands were described on plants from the dump-field area (e.g. on *Picea abies*, *Pinus sylvestris*, *Larix decidua*, *Salix caprea*, *Betula pendula* etc.). Site influence was statistically tested on size of *Betula pendula* leaves. For statistical analysis the STATISTICA 7.0 statistical package was used (StatSoft Inc., 2004). ANOVA test enabled evaluate differences in width of leave bases among individual plants on dump-field surface with those from reference area. Measuring of 50 individuals was realised from each locality. For each individual plant were measured on 10 leaves leave-blade bases from 8 pieces of one year old branches. Main descriptive characteristics of investigated set of plants both from dump-field and from reference area are presented in Table 6.

Table 6. Main descriptive characteristics of investigated set of plants

	Dump-field Podlipa	Reference area
Sample size	400	400
Arithmetic mean	2.92	5.76
Median	2.87	5.74
Mode	2.61	5.71
Variance / dispersion	0.237	0.501
Standard deviation	0.487	0.708
Minimum	1.62	3.64
Maximum	5.23	11.05
Range	3.61	7.41

Result of Least Significant Difference (LSD) shows statistically significant difference between the wide of base of birch leaves (it means also total size of leaves) from individuals of Cu-dump field and reference locality with $p < 0.05$ (Table 7).

Table 7. Result of LSD with $p < 0.05$

Source	Sum of squares	Degrees of freedom	Mean square	F-ratio	p-value
Between groups	1618.82	1	1618.82	4386.38	0.000
Within groups	294.506	798	0.369		
Total	1913.32	799			

4. Discussion

Heavy metal distribution at the investigated dump-fields at Ľubietová reflect the geochemical behavior of the elements, depending on their content, solubility, migration potential and sorption properties (Cataldo & Wildung, 1978). Hydrolysable metals (e.g., Ni, Cd) or metals forming insoluble precipitates with S or P on entering the soil in soluble forms may be expected to be rapidly insolubilised at the near neutral pH of most soils due to hydrolysis on dilution and subsequent precipitation on, or reaction with particle surfaces. Certain elements (e.g. Fe) may also form precipitates with S or P (Routson & Wildung, 1969). The soil physicochemical parameters, most important in influencing the solubility of metals include: composition of solution, Eh, and pH; type and density of charge on soil colloids; and reactive surface area (Keeney & Wildung, 1977).

The main Cu sources at the dump-fields are sulphides (tetrahedrite and chalcopyrite) and secondary Cu-minerals (libethenite, langite, brochantite, pseudomalachite, malachite and azurite). Cu released to the solution during weathering processes, contaminates aqueous medium.

Mobility of the majority of heavy metals is in the nature mostly determined by their sorption on natural sorbents, which are mainly represented by clay minerals (Missana et al., 2008).

Kaolinite has been used as a good sorbent for most heavy metals (Wahba & Zaghloul, 2007). Cu, Pb, Zn and Cd show favourable sorption on smectite and Pb also on illite surface (Rybicka et al., 1995). The uptake of Pb and Cu on illite and smectite is usually very fast. Kinetics of Zn, Ni and Cd sorption on illite and smectite are not so efficient. Mg, Fe and Al sorption on clay minerals is more efficient at higher pH. It is caused by absence of free H^+ ions and by increase of negative charge on clay minerals surface (Kish & Hassan, 1973). $pH_{(H_2O)}$ of the technogenous sediments at Podlipa dump-fields ranges from 4,21 to 7,93 ($pH_{(KCl)}$ 4.00–7.34), thus conditions for Cu, Pb, Zn and Cd adsorption on clay minerals are not the best but also not inefficient.

Thorium contents in clay minerals are amazingly lower than in the dump-sediment. It means that the Th is during the maceration washed out from the clays. This trend is noticeable because U is generally considered to be more mobile as Th (Polanski & Smulikowski, 1978). The better mobility of U at the Ľubietová deposit caused that while the content of Th in soil is several times higher than the content of U, while in plants (Andráš et al., 2007) is the Th/U rate in consequence of better U mobility approximately identical (about 1 : 1).

Acidity is mainly up to the geochemical behavior (weathering) of particular minerals (mainly pyrite). The calculation of the acid mine drainage water (AMD) formation potential (neutralisation potential, total acidity production, net neutralisation potential) is also discussed. The value of the net neutralisation potential (NPP) and the NP : AP ratio show that the potential of the acid mine drainage water formation is very limited (NPP = 1,42; NP : AP = 1,72) and the environmental risk is negligible.

Lack of carbonates at Ľubietová deposit causes that in 5 cases among 12 studied samples, are the NNP values negative (neutralisation matter is entirely absent) and two values (samples A-3 and A-11) are very low (7.4 and 20.1; Table 4). NNP values from -20 to 20 (kg $CaCO_3 \cdot t^{-1}$ of dump material) are possible to account as a *"scale of uncertainty"* sensu US EPA methodics (Lintnerová & Majerčík, 2005) from viewpoint of the acidity production, because there is no unambiguous forecast if the AMD will be produced. In spite of this fact results of our study suggest that at Reiner and Podlipa is the assumption of AMD production very limited.

Mining dump surrounding is characterized by no or very poor and dry soil cover with lack of soil nutrinents, minimum of water combined with intensive evaporation, strong solifluction patterns and high heavy metal content. The plants growing at such as habitats are usually tolerant to a high metal content, characterized by high vitality and typical for xerothermic conditions.

As a symptom of heavy metal contamination is possible mention violet to red stigmas and necrosis on leaves and stems of vascular plants, nanic growth (Andráš et al., 2007; Chaves et al., 2011), reduction of roots (Banásová, 1976), chlorosis of leaves with green veining, growth stagnancy (Kopponen et al., 2001), drying of young branches and decrease of leaves size (Pulford & Watson, 2003), ultrastructural effects (Ouzounidou et al., 1992).

Toxicity symptoms in plants from mining dumps (tendency for mould diseases, growth decrease, deformations of cell organelles and plant tissues, chlorosis of leaves) show sometimes similar features as influence of lack of essential elements. If the soil contains very high content of copper, we can in the plants see lack of Fe as a consequence of immobilisation of Fe by Cu. As a result of Fe-lack we can see chlorosis of leaves, which could be caused also by superabundance of Zn (Kopponen et al., 2001). To state the unambiguous specification of toxicity reasons is necessary to have information about the heavy metal content in soil-plant system.

Betula pendula is the tree which represents the most important dynamic element on mining dumps and is considered as a focus of initial stages of vegetation due to litter formation, shadow and leeward (Banásová, 1976). *Betula pendula* is a suitable bioindicator of air pollution in urban areas (Samecka et al., 2009; Petrova, 2011) and it is used as a model organism in several studies focused on heavy metal influence on growth, annual additions and reproductive effort (Samecka et al., 2009; Franiel & Babczynska, 2011). Our study also confirmed decrease of birch leave blades caused by heavy metal content.

5. Conclusions

The soil and technogenous sediments contamination is very irregular. It depends on the mineralogical composition of ores, on the original concentration of the mentioned metals in the technogenic sediments of the spoil dumps, and also on their migration abilities and sorption properties. Bilateral correlation of metals seems to be influenced by incorporation of elements to various metastable phases as well as by present natural sorbents. This proces is controlled both by oxidation state and by ionic radius of elements.

The surface water (and drainage water) as well as the groundwater water are substantially contaminated predominantly by Cu, Fe, As and Sb. The content of the most dangerous contaminants: As^{3+}, As^{5+}, Sb^{3+} and Sb^{5+} don't pose acute risk. The only risk poses the spring of the mineral water Linhart because of the high radioactivity and high Fe, Cu, Cd and Pb contents.

The present natural sorbents are predominantly the clay minerals (illite, muscovite, caolinite, smectite) and hydrogoethite. The clay minerals are good sorbents of V, Cr, Ti, W, Zr, Nb, Ta a Th and at the hydrogoethite of Cu, Zn, Mo, Mn, Mg, (± Fe, Cd, Co, Ca). In the case of the Fe, As, Sb, Ag, Pb, Zn, Mn, Mo, Bi, U was proved also the free sorption capacity.

The paste or rinse pH of sediments measured in distilled H_2O is around 5.3 and only very few samples account acid values (< 5.0). Only several few samples show markedly acid reaction. The acidity production (AP) vary from 0.625 to 10.31 (in average 3.7) and the neutralisation potential (NP $CaCO_3$) from 0.66 to 12.71 $kg \cdot t^{-1}$ (in average ca 27.1 $kg \cdot t^{-1}$ $CaCO_3$). The value of the net neutralisation potential (NNP) and the NP : AP ratio show that the potential of the acid mine drainage water formation is very limited (NNP = 1.42; NP : AP = 1.72) and the environmental risk is negligible.

The concentrations of the heavy metals in plant tissues decrease seriately in rate: Fe, Zn, Pb and Cu. The highest concentrations of heavy metals are in roots, than in leaves and stems and the lowest concentrations are in flowers, seeds and in fruits. The plant tissues from the dump-field are heavily damaged. The results of the research document the plant defence reactions under the influence of stress factors at the dump sites (absence of soil and water, the heavy metal contamination, mobility of the cohesionless slope material).

Botanic research indicated that the specific conditions at the dump-fields have evident influence on plants. Some damage spots are visible even by naked eye but some of them were proved also statistically, by histological study of *Pinus sylvestris* and by evaluation of morbid changes of *Acetosella vulgaris*.

The Ľubietová-Podlipa dump-field dispose by certain degree of "*self-cleaning ability*". Important part of the heavy metals and contaminants is fixed in porous dump-material, Fe-hydroxides and in clay minerals (mainly illite, caolinite, smectite and chlorite group), which show still an important free sorption capacity. The ability of the drainage water precipitate cementation copper (as well as Sb and probably also other heavy metals as As) on the iron surface give possibility to realize Fe^0-barrier for elimination of heavy metals from the groundwater and drainage water.

Acknowledgements

This work was supported by grant contracts VEGA 2-0065-11 and APVV-0663-10. We would like to thank Ján Ostrolúcky and Ján Tomaškin for help with field work and statistical analyses. This paper has been elaborated in the framework of the project Opportunity for young researchers, reg. no. CZ.1.07/2.3.00/30.0016, supported by Operational Programme Education for Competitiveness and co-financed by the European Social Fund and the state budget of the Czech Republic.

References

Adams, J. P., Adeli, A., Hsu, C. Y., Harkess, R. L., Page, G. P., Schultz, E. B., & Yuceer, C. (2011). Poplar maintains zinc homeostasis with heavy metal genes HMA4 and PCS1. *Journal of experimental botany, 62*(11), 3737-3752. http://dx.doi.org/10.1093/jxb/err025

Andráš, P., Mamoňová, M., Ladomerský, J., Turisová, I., Lichý A., & Rusková, J. (2007). Influence of the dump sites on development of selected plant tissues at the Ľubietová area (Slovakia). *Acta Facultatis Ecologiae, 16*(1), 147-158.

Baker, A. J. M., Ernst, W. H. O., van der Ent, A., Malaisse, F., & Ginocchio, R. (2010). Metallophytes: the unique biological resource, its ecology and conservational status in Europe, central Africa and Latin America. In L. C. Batty & K. B. Hallberg (Eds.), *Ecology of industrial pollution* (pp. 7-40). Cambridge: Cambridge University Press. http://dx.doi.org/10.1017/CBO9780511805561.003

Banásová, V. (1976). Vegetácia medených a antimónových háld (Vegetation of copper and antimony mine heaps). *Biologické Práce, 22*, 1-109.

Bartzas, G., Komnitsas, K., & Paspaliaris, I. (2006) Laboratory evaluation of Fe^0 barriers to treat acidic leachates. *Minerals Engineering, 19*(5), 505-514. http://dx.doi.org/10.1016/j.mineng.2005.09.032

Bergfest, A. (1951). *Baníctvo v Ľubietovej na medenú rudu.* Banská Štiavnica: Ústredný banský archív pre Slovensko-Central Mining Archive.

Cataldo, D. A., & Wildung, R. E. (1978). Soil and plant factors influencing the accumulation of heavy metals by plants. *Environmental Health Perspectives, 27*, 149-159. http://dx.doi.org/10.1289/ehp.7827149

Chaves, L. H. G., Estrela, M. A., & de Souza, R. S. (2011). Effect on plant growth and heavy metal accumulation by sunflower. *Journal of Phytology, 3*(12), 4-9.

Ebner, F., Pamić, J., Kovács, S., Szederkenyi, T., Vai, G. B., Venturini, C., ... Mioč, P. (2004). Variscan Preflysch (Devonian—Early Carboniferous) environments. *Tectonostratigraphic Terrane and Paleoenvironment Maps of the Circum-Pannonian Region, 1*(2,500,000).

Fairthorne, G., Fornasiero, D., & Ralston, J. (1997). Effect of oxidation on the collectorless flotation of chalcopyrite. *International Journal of the Processes, 49*(1-2), 31-48. http://dx.doi.org/10.1016/S0301-7516(96)00039-7

Franiel, I., & Babczyńska, A. (2011). The Growth and Reproductive Effort of *Betula pendula* Roth. In a Heavy-Metals Polluted Area. *Polish Journal of Environmental Studies, 20*(4), 1097-1101.

Greenwood, N. N., & Earnshaw, A. (1990). *Chemie der Elemente.* Würzburg: VCH Verlagsgesellschaft mbH.

Keeney, D. R., & Wildung, R. E. (1977). Chemical properties of soils. In L. Elliott & R. J. Stevenson (Eds.), *Soils for Management of Organic Wastes and Waste* (pp. 75-100). Madison, Wisconsin: American Society of Agronomy.

Kishk, F. M., & Hassan, M. N. (1973). Sorption and desorption of copper by and from clay minerals. *Plant Soil, 39*(3), 497-505. http://dx.doi.org/10.1007/BF00264168

Koděra, M. (1990). *Topografická mineralógia 2*. Bratislava: Veda, SAV.

Kopponen, P., Utriainen, M., Lukkari, K., Suntioinen, S., Kärenlampi L., & Kärenlampi, S. (2001). Clonal differences in copper and zinc tolerance of birch in metal-supplemented soils. *Environmental Pollution, 112*(1), 89-97. http://dx.doi.org/10.1016/S0269-7491(00)00096-8

Lintnerová, O., & Majerčík, R. (2005). Neutralizačný potenciál sulfidického odkaliska Lintich pri Banskej Štiavnici–metodika a predbežné hodnotenie (Neutralisation potential of tailing of sulphide deposit at Lintich near Banská Štiavnica). *Mineralia Slovaca, 37*(4), 17-28.

Manning, B. A., & Goldberg, S. (1997) Adsorption and stability of Arsenic(III) at the clay mineral-water interface. *Environmental Science & Technology, 31*(7), 2005-2011. http://dx.doi.org/10.1021/es9608104

Missana, T., Garcia-Guttierez, M., Alonso, U. (2008). Sorption of strontium onto illite/smectite mixed clays, *Physics and Chemistry of the Earth, 33*(Sup. l.1), 156-162.

Morin, K. A., & Hutt, N. M. (1997). Environmental geochemistry of minesite drainage: Practical Tudory and case studies. Vancouver: MDAG Publishing.

Ouzounidou, G., Eleftheriou, E. P., & Karataglis, S. (1992). Ecophysical and ultrastructural effects of copper in Thlaspi ochroleucum (Cruciferae). *Canadian Journal of Botany, 70*(5), 947-957. http://dx.doi.org/10.1139/b92-119

Petrova, S. T. (2011). Biomonitoring Study of Air Pollution with *Betula pendula* Roth. from Plovdiv, Bulgaria. *Ecologia Balkanica, 3*(1), 1-10.

Polański, A., & Smulikowski, K. (1978). *Geochémia*. Bratislava: Slovenské pedagogické nakladateľstvo.

Pulford, I. D., & Watson, C. (2003). Phytoremediation of heavy metal-contaminated land by trees – a review. *Environment International, 29*(4), 529-540. http://dx.doi.org/10.1016/S0160-4120(02)00152-6

Routson, R. C., & Wildung, R. E. (1969). Ultimate disposal of wastes to soil. *Chemical Engineering Progress, Symp. Ser., 65*(97), 19-25.

Rybicka, E. H., Calmano, W., & Breeger, A. (1995). Heavy metals sorption/desorption on competing clay minerals; an experimental study. *Applied Clay Science, 9*(5), 369-381. http://dx.doi.org/10.1016/0169-1317(94)00030-T

Ryu, J., Gao, S., Dahlgren, R. A., & Ziernberg, R. A. (2002) Arsenic distribution, speciation and solubility in shallow groundwater of Owens Dry Lake, California. *Geochimica et Cosmochimica Acta, 66*(17), 2981-2994. http://dx.doi.org/10.1016/S0016-7037(02)00897-9

Samecka-Cymerman, A., Kolon, K., & Kempers, A. J. (2009). Short shoots of *Betula pendula* Roth. as bioindicators of urban environmental pollution in Wrocław (Poland). *Trees - Structure and Function, 23*(5), 923-929. http://dx.doi.org/10.1007/s00468-009-0334-z

Sobek, A. A., Schuller, W. A., Freeman, J. R, & Smith, R. M. (1978). *Field and laboratory methods applicable to overburden and minesoils*. EPA 600/2-78-054.

StatSoft, I. N. C. (2004). STATISTICA (data analysis software system), version 7. *Computer software*. Retrieved from www.statsoft. com

Vink, B. W. (1966). Stability relations of antimony and arsenic compounds in the light of revised and extended Eh-pH diagrams. *Chemical Geology, 130*(1-2), 21-30. http://dx.doi.org/10.1016/0009-2541(95)00183-2

Wahba, M. M., & Zaghloul, A. M. (2007). Adsorption Characteristics of Some Heavy Metals by Some Soil Minerals. *Journal of Applied Sciences Research, 3*(6), 421-426.

5

Physicochemical Changes in the Quality of Surface Water due to Sewage Discharge in Ibadan, South-Western Nigeria

Oluseyi E. Ewemoje[1] & Samuel O. Ihuoma[1]

[1] Department of Agricultural and Environmental Engineering, University of Ibadan, Ibadan, Nigeria

Correspondence: Oluseyi E. Ewemoje, Department of Agricultural and Environmental Engineering, University of Ibadan, Ibadan, Nigeria. E-mail: seyiajayi2@yahoo.com

Abstract

Sewage discharge is known to degrade the quality of receiving water bodies. This study assesses the impact of black water discharge on the physico-chemical parameters of River Zik in the University of Ibadan. Water samples were collected from five sampling sites along the stream located at progressive distances from the discharge point. Sampling was done three times over a period of three months (May to July 2012). The physico-chemical parameters tested were: pH, Biochemical Oxygen Demand (BOD_5), Dissolved Oxygen (DO), Electrical Conductivity (EC), Total Suspended Solids (TSS) and Nitrate. The overall mean values of the measured parameters were as follows: BOD_5 (381.1 mg/L); DO (3.9 mg/L); SS (1825.4 mg/L); pH (6.1); EC (618.5 μs/cm) and Nitrates (59.8 mg/L). The highest concentrations of BOD_5, SS, EC and Nitrates were obtained at the point of sewage discharge into the stream. One-way ANOVA showed significant deviation from WHO standards for BOD_5, SS, DO, EC and Nitrates ($p < 0.05$). This study showed that sewage discharge into River Zik have seriously contributed to the pollution of the stream to levels which pose health and environmental hazards to those using it downstream for domestic and agricultural purposes. This environmental hazards has been attributed to the little or near none existence of regulatory bodies responsible for regulating the strenght of black water discharge into sewers and/or recieving water bodies.

Keywords: sewage, receiving water body, environmental hazards, down-stream pollution and South-Western Nigeria

1. Introduction

Wastewater is referred to as any water that has been previously used by domestic or industrial activity and because of the usage now contains waste products. This waste products due to anthropogenic influence and natural processes such as precipitation, weathering and sediment transportation affects the water quality (Qadir, Malik, & Husain, 2008) which comprises of discarded domestic solid or liquid, industrial, agricultural and commercial gaseous chemical and radioactive wastes (Ekhaise & Omavwoya, 2008; Al-Ghamdi, 2011). The major components of water pollution are wastewater discharged from sewer systems and the industries. Wastewater from sewer systems contribute to increased oxygen demand and nutrient loading of the water bodies thereby promoting toxic algal blooms which usually leads to a destabilized aquatic ecosystem (Morrison, Fatoki, Persson, & Ekberg, 2001; Igbinosa & Okoh, 2009).

Sewage discharge leading to water pollution has become an issue of considerable public and scientific concerns in the light of evidence of their extreme toxicity to human health and ecosystems. Pollution is caused when a change in the physical, chemical or biological condition in the environment harmfully affect quality of human life including other animals' life and plant (Okoye, Enemuoh, & Ogunjiofor, 2002). Industrial, sewage, municipal wastes are been continuously added to water bodies hence affecting the physicochemical quality of water making them unfit for use of livestock and other organisms (Pandey, 2003).

According to previous research works (Qadir et al., 2008; Pandey, 2006); uncontrolled domestic wastewater discharge into the streams without any form of treatment has resulted in eutrophication of the water bodies as evidence by substantial algal bloom; dissolve oxygen depletion in the subsurface water leads to large fish kill and other oxygen requiring organisms. Sewage discharge into the environment with enhanced concentration of nutrients, sediments and toxic substances may have a serious negative impact on the quality and life forms of the receiving water body when discharge untreated or partially treated (Miller & Siemmens, 2003; Schulz & Howe,

2003).

Water associated diseases which are associated with ingestion of contaminated water with human or animal faeces are the major source of faecal microorganisms, including pathogens and microbial intestinal infections such as cholera, typhoid fever and bacillary dysentery (Cabral, 2010). These diseases caused by bacteria, viruses and protozoa are the most common health risks associated with contaminated water sources (WHO, 2008). These contain wide varieties of viruses, bacteria, and protozoa that may get washed into drinking water supplies or receiving water bodies (Kris, 2007). Virus concentrations present in raw water receiving fecal matter from humans are often high, although these viruses cannot reproduce in water; however, they are still capable of causing diseases when ingested even at low doses (Okoh, Sibanda, & Gusha, 2010). Many microbial pathogens in wastewater can cause chronic diseases with costly short and long-term effects, such as degenerative heart disease and stomach ulcer (Akpor & Muchie, 2011). Bacteria cause a wide range of infections, such as diarrhea, dysentery, skin and tissue infections. Disease-causing bacteria found in water include different types of bacteria, such as E. coli, Listeria, Salmonella, Leptospirosis, Vibrio and Campylobacter (Absar, 2005). The most common and widespread health risks associated with drinking poor quality water in developing countries are of biological origin and diarrhoeal disease globally has been attributed to unsafe water, sanitation and water hygiene (Suthar, Chhimpa, & Singh, 2009). Other impacts of discharging untreated or inadequately treated wastewater into the environment according to Okoh, Odjadjare, Igbinosa and Osode (2007) include increased nutrient levels (eutrophication), often leading to algal blooms; depleted dissolved oxygen, sometimes resulting in fish kills; destruction of aquatic habitats with sedimentation, debris, and increased water flow; and acute and chronic toxicity to aquatic life from chemical contaminants, as well as bio-accumulation and bio-magnification of chemicals in the food chain. This study was therefore conducted to assess the impact of sewage discharge on the physico-chemical parameters of river Zik in university of Ibadan, Nigeria.

2. Methodology

2.1 Study Site

This study was carried out in university of Ibadan, Ibadan which is located in south-western Nigeria between longitude 3°58'E and latitude 7°22'N. The altitude generally ranges from 15 to 21 m above mean sea level (Oyediran & Adeyemi, 2012). The stream flows from Sango in Ibadan town through the institution and empties in Awba dam of the university. The dam is used for fishing and irrigation while, water is abstracted from the dam by the university water treatment plant for treatment and distribution to the residence on campus.

2.2 Sampling Design

Water samples were collected at five different points along the stream using random grab sampling. Sampling was done three times from each of the points and all samples were collected in triplicate to improve reliability of data. Water samples were collected from the mid-width of the stream using one-litre plastic bottles that had previously been cleaned, soaked in 10% nitric acid and rinsed thrice with distilled water. Three one-litre samples were collected at each of the sampling points designated A – E. The full description of sampling locations is shown in Table 1.

Table 1. Sampling location description

Designation	Sampling points
A	10 m upstream with respect to sewage discharge point.
B	Point of discharge.
C	10m downstream from point of discharge.
D	50m downstream from point of discharge.
E	100m downstream from point of discharge.

2.3 Physico-Chemical Properties of the Wastewater Samples

Samples were analyzed for the following physico-chemical parameters: Hydrogen ion Concentration, Temperature, Total Dissolved Solid (TDS), Dissolved Oxygen (DO), Biochemical Oxygen Demand (BOD5), Chemical Oxygen Demand (COD) and Electrical Conductivity. The pH value of the samples were determined with a pH meter (Unicam 9450, Orion model No. 91-02). Temperature was measured with mercury thermometer immediately after sample collection. Gravimetric method involving filtration and evaporation were used to measure the total dissolved solids. Methods recommended by APHA (1998) were followed for the measurement of DO, BOD_5 and COD. The Electrical Conductivity was determined using a conductivity meter (Metrohm 640, Switzerland).

2.4 Statistics

The results of laboratory analysis were subjected to data analysis using SPSS, version 12. To analyze changes in the levels of BOD_5, TSS, DO, pH, EC and Nitrates that might be attributed to sewage discharge into River Zik. One-way Analysis of Variance (ANOVA) at 95% level of significance was used.

3. Results and Discussion

A summary of the results collected over a period of three months, between May to July 2012 is presented in the following set of tables which also compare each of the parameters with the acceptable levels. A discussion of each parameter follows the tables.

At point A, 10m upstream, the BOD_5 value was 36 ± 1.8 mg L^{-1} (Table 2). The BOD concentration went higher to 955 ± 16.7 mg L^{-1} at the point B. This was because of the sewage discharged on the stream at this point and therefore not completely mixed. At point C, 10m downstream from the point of discharge, the mean value of BOD_5 was reduced to 733 ± 35.2 mg L^{-1}. This can partially be attributed to dilution due to mixing and partially as a result of settling along the stream course and dilution. Also, some part of BOD_5 may decrease due to microbial degradation during course of flow from B to C. However, the time required for water to travel this distance is too small to get any significant degradation. The flow of the water in the stream may increase the amount of dissolved oxygen in water that subsequently increased the microbial degradation of organic matter. At points D (50m upstream) and E (100m upstream), the mean BOD_5 values further decreased to 416 ± 36.3 and 237 ± 15.8 mg L^{-1} respectively. This, again, may be due to extensive dilution occurring in the stream during flowing of the effluent as it moves away from point A. Nevertheless, the mean values of BOD_5 at points D and E remained higher and may be partly due to other non-point pollution sources or that the microbial load far exceeds the self-purification capacity of the stream.

Table 2. Biochemical Oxygen Demand (BOD_5) in River Zik (Means and Standard errors of the means) compared with WHO standard

Sample station	BOD in the Stream (mg/L)	Acceptable WHO levels	WHO standards exceeded by (mg/L)
A	36 ± 1.8	20	16
B	955 ± 16.7	20	935
C	733 ± 35.2	20	713
D	416 ± 36.3	20	396
E	237 ± 15.8	20	217

Results (Table 2) from all the sampling points indicate that BOD_5 levels far exceed the acceptable levels of 20 mg L^{-1} as stipulated by World Health Organization (WHO, 1984). The low levels of BOD_5 at point A indicate that the stream is not heavily polluted before the discharge of raw sewage. There was, however, a marked increase at the point of discharge indicating a substantial impact and degradation of water quality.

The DO concentration from the study (Table 3) ranged from 1.6 ± 0.1 to 5.8 ± 0.2 mg L^{-1}. DO is the measure of the degree of pollution by organic matter, the destruction of organic substance as well as self-purification of the water bodies. It reflects interaction with the overlaying air because oxygen from the atmosphere is dissolved in the water (Chiras, 1998) and it is one of the most significant tests for measuring the quality of water. The standard for sustaining aquatic life is stipulated to be 5 mg L^{-1} (Horne & Goldman, 1994). Concentration below 2 mg L^{-1} adversely affects aquatic and biological life while the concentration below 2 mg L^{-1} may lead to death of fish. The lowest mean value of 1.6 ± 0.1 mg L^{-1} was detected at B (a point where the effluent is discharged into the stream). There is a decrease in concentration from 3.6 ± 0.1 mg L^{-1} found at C. The low DO concentration at point B could be due to high organic load of BOD_5 and suspended solid values (Table 2). At point C, the mean DO level increased to 3.6 ± 0.1 mg L^{-1} probably due to mixing and re-aeration along the stream.

Table 3. Dissolved Oxygen (DO) in River Zik (Means and Standard errors of the means) compared with standard to support aquatic life

Sample station	DO in the Stream (mg/L)	Standard for Sustaining Aquatic Life (mg/L)	Standard for Sustaining Aquatic life exceeded by (mg/L)
A	4.5 ± 0.2	5	0.5
B	1.6 ± 0.1	5	3.4
C	3.6 ± 0.1	5	1.4
D	4.1 ± 0.1	5	0.9
E	5.8 ± 0.2	5	Not exceeded

Table 4. Suspended Solids (SS) in River Zik (Means and Standard errors of the means) compared with WHO standard

Sample station	SS in the Stream (mg/L)	Acceptable WHO levels (mg/L)	WHO standards exceeded by (mg/L)
A	1510.6 ± 217.2	30	1480.6
B	2476.9 ± 21.2	30	2446.9
C	1838.8 ± 81.4	30	1808.8
D	1766.9 ± 48.0	30	1736.9
E	1534.0 ± 33.6	30	1504

Table 5. Electrical Conductivity (EC) in River Zik (Means and Standard errors of the means) compared with WHO standard

Sample station	EC in the Stream (μS/cm)	Acceptable WHO levels	WHO standards exceeded by (mg/L)
A	435.5 ± 3.2	400	35.5
B	905.3 ± 9.4	400	505.3
C	741.1 ± 5.1	400	341.1
D	588.0 ± 3.2	400	188
E	422.7 ± 3.7	400	22.7

The mean value of DO concentration continued to improve down the stream to 4.1 ± 0.1 and 5.8 ± 0.2 mg L^{-1} at points D and E, respectively. This could be attributed to both the flow and recovery capacity of the stream. At C, it may be suggested that the stream recovered from the organic load and could have been a better status to support aquatic life.

Electrical conductivity (EC) values (Table 5) ranged from 422.7 ± 3.7 to 905.3 ± 9.4 μS cm^{-1}. WHO recommends 400 μS cm^{-1}. At point A, an average of 435.5 ± 3.2 μS cm^{-1} was recorded. The high value could be due to the discharge of gray water few meters upstream, leading to concentration of salts from detergents. The value rose to 905.3 ± 9.4 μS cm^{-1} at point B as a result of the sewage discharge. At point C, the recorded conductance value was 741.1 ± 5.1 μS cm^{-1} and decreased to 588.0 ± 3.2 and 422.7 ± 3.7 μS cm^{-1} at points D and E respectively.

The reduction may be due to little amounts of dissolve solids in water due to dilution. Electrical conductivity is used to indicate the dissolved solids in water because the concentration of ionic species determines the conduction of current in an electrolyte (Hayashi, 2004). The high value of electrical conductivity therefore suggest that river Zik has a considerable loading of dissolved salts, and is unsuitable for irrigation, having exceeded the minimum acceptable levels as stipulated by WHO.

Hydrogen ion concentration or pH is the indicator of acidity or alkalinity of water. It is a measure of the effective concentration (activity) of hydrogen ions in water. Water having a pH range of 6.5–8.5 will generally support a good number of aquatic species. Only a few species can tolerate pH values lower than 5 or greater than 9 (Harrison, 1999). The mean values of pH obtained (Table 6) ranged from 6.0 ± 0.2 to 6.2 ± 0.1. These pH values were slightly below the recommended ranges (6.5–9.5) as stipulated by WHO. However, it falls within the EU standard pH limits of 6.0 to 9.0 for protection of fisheries and aquatic life (Chapman, 1996).

The most highly oxidized form of nitrogen compounds is commonly present in surface and groundwater because it is the end product of aerobic decomposition of organic nitrogenous matter. Unpolluted natural waters usually contain only minute amounts of nitrate (Jaji, Bamgbose, Odukoya, & Arowolo, 2007). Usually, nitrate levels of streams polluted by human and animal wastes are high. In this study, the nitrate-N concentrations was generally high and ranged between 17.6 ± 0.2 to 97.3 ± 1.2 mg/L. It is important to note that nitrate level in the stream could be a source of eutrophication for receiving water. The obtained values exceeded the recommended limit for WHO except for points A and E, 10 m upstream and 100 m downstream from the point of sewage discharge respectively. The nutrient levels in the upstream discharge point of the receiving water may be as a result of diffuse sources from settlement and agricultural runoff, while the reduced level of nitrate in the downstream point could be as a result of conversion of nitrates to nitrites along the length of the stream.

Table 6. The pH in River Zik (Means and Standard errors of the means) compared with WHO standard

Sampling stations	pH in the Stream (mg/L)	WHO Recommended Range
A	6.1 ± 0.2	6.5–9.5
B	6.0 ± 0.2	6.5–9.5
C	6.0 ± 0.2	6.5–9.5
D	6.1 ± 0.1	6.5–9.5
E	6.2 ± 0.1	6.5–9.5

Table 7. Nitrate in River Zik (Means and Standard errors of the means) compared with WHO standard

Sampling stations	Nitrate in the Stream (mg/L)	Acceptable WHO levels	WHO Maximum Limit exceeded by (mg/L)
A	17.6 ± 0.2	25 - 50	Not exceeded
B	97.3 ± 1.2	25 - 50	47.3
C	85.9 ± 0.3	25 - 50	35.9
D	65.8 ± 0.6	25 - 50	15.8
E	32.6 ± 0.8	25 - 50	Not exceeded

One way ANOVA test showed that only BOD_5, SS, DO, EC and Nitrates had significant difference ($P < 0.05$). This implies that BOD_5, DO, SS, EC and Nitrates were the parameters of water quality of the stream that had been significantly affected by pollution.

Similar studies done in Tanjero River (Mustapha, 2002) and in several other streams and rivers found that sewage discharge in surface water is contributing greatly to the degradation of ecosystem health.

4. Conclusion

This study showed that most of the parameters assessed in river Zik are above limits recommended by WHO. The study showed that although the stream was already polluted from upstream activities, there was a significant increase in the levels of BOD_5, SS, DO, EC and Nitrates at the raw sewage discharge point, thereby, contributing to the pollution of river Zik and endangering ecosystem and the health of the people who rely on the stream for livelihood water source. The management of sewage wastewater should be improved to minimize danger to the environment and to people. Relevant authorities should embark on regular monitoring activities of streams and rivers to ensure the safety of human and animal population and the environment. It is recommended that further research should be carried out in the low flow season where minimum or near zero flows are observed as the river is ephemeral in nature.

References

Absar, A. K. (2005). Water and Wastewater Properties and Characteristics. In J. H. Lehr & J. Keeley (Eds.), *Domestic, Municipal and Industrial Water Supply and Waste Disposal* (pp. 903-905), New Jersey: John Wiley and Sons, Inc.

Akpor, O. B., & Muchie, B. (2013). Environmental and public health implications of wastewater quality. *African Journal of Biotechnology, 10*(13), 2379-2387.

Al-Ghamdi, A. Y. (2011). Review on Hospital Wastes and its Possible Treatments. *EgyptAcad. J. Biolog. Sci., 3*(1), 55-62.

APHA. (1998). *Standard Methods of Examination of Water and Wastewater* (20th ed.). Washington D.C.

Cabral, J. P. S. (2010). Water Microbiology. Bacterial Pathogens and Water. *Int. J. Environ. Res. Public Health, 7*, 3657-3703. http://dx.doi.org/10.3390/ijerph7103657

Chapman, D. V. (Ed.). (1996). *Water quality assessments: a guide to the use of biota, sediments and water in environmental monitoring* (p. 626). London: E & Fn Spon. http://dx.doi.org/10.4324/NOE0419216001

Chiras, D. D. (1998). *Environmental Science. A Systematic Approach to Sustainable Development* (5th ed). Washington D.C: Wadsworth Publishing Co.

Ekhaise, F. O., & Omavwoya, B. P. (2008). Influence of Hospital Wastewater Discharged from University of Benin Teaching Hospital (UBTH), Benin City on its Receiving Environment. *American-Eurasian J. Agric. & Environ. Sci., 4*(4), 484-488.

Harrison, R. M. (Ed.). (1999). *Understanding our environment: An introduction to environmental chemistry and pollution*. Royal Society of chemistry.

Hayashi, M. (2004). Temperature-Electrical Conductivity Relation of Water for Environmental Monitoring and Geophysical Data Inversion. *Environ. Monit. Assess., 96*, 199-128. http://dx.doi.org/10.1023/B:EMAS.0000031719.83065.68

Horne, A. J., & Goldman, C. R. (1994). *Limnology* (2nd ed., p. 480). New York: McGraw-Hill.

Igbinosa, E. O., & Okoh, A. I. (2009). Impact of discharge wastewater effluents on the physico-chemical qualities of a receiving watershed in a typical rural community. *International Journal of Environmental Science & Technology, 6*(2), 175-182. http://dx.doi.org/10.1007/BF03327619

Jaji, M. O., Bamgbose, O., Odukoya, O. O., & Arowolo, T. A. (2007). Water quality assessment of Ogun River, south west Nigeria. *Environmental monitoring and assessment, 133*(1-3), 473-482. http://dx.doi.org/10.1007/s10661-006-9602-1

Kris, M. (2007). *Wastewater Pollution in China*. Retrieved June 16, 2011, from http: www.dbc.uci/wsu stain/suscoasts/krismin.html

Miller, G. L. (1959). "Use of Dinitrosalicylic Acid Reagent for Determination of Reducing Sugar". *Anal. Chem., 31*, 426-428. http://dx.doi.org/10.1021/ac60147a030

Morrison, G., Fatoki, O. S., Persson, L., & Ekberg, A. (2001). Assessment of the impact of point source pollution from the Keiskammahoek Sewage Treatment Plant on the Keiskamma River-pH, electrical conductivity, oxygen-demanding substance (COD) and nutrients. *Water SA, 27*(4), 475-480. http://dx.doi.org/10.4314/wsa.v27i4.4960

Mustafa, O. M. (2006). Impact of Sewage Wastewater on the Environment of Tanjero River and Its Basin within Sulaimani City/NE-Iraq. *Geology MSc Thesis. Baghdad: College of Science, University of Baghdad.*

Okoh, A. I., Odjadjare, E. E., Igbinosa, E. O., & Osode, A. N. (2007). Wastewater treatment plants as a source of microbial pathogens in receiving watersheds. *African Journal of Biotechnology, 6*(25).

Okoh, A. I., Sibanda, T., & Gusha, S. S. (2010). Inadequately treated wastewater as a source of human enteric viruses in the environment. *International journal of environmental research and public health, 7*(6), 2620-2637. http://dx.doi.org/10.3390/ijerph7062620

Okoye, P. A. C., Enemuoh, R. E., & Ogunjiofor, J. C. (2002). Traces of heavy metals in Marine crabs. *J. Chem. Soc. Nig, 27*(1), 76-77.

Oyediran, I. A., & Adeyemi, G. O. (2012). Geochemical Assessment of a Proposed Landfill in Ibadan, Southwestern Nigeria". *Pacific Journal of Science and Technology, 13*(1), 640-651.

Pandey. (2003). Trends in Eutrophication Research and Control. *Hydrol. Proc., 10*(2), 131-295.

Pandey, S. N. (2006). Accumulation of heavy metals (Cd, Cr, Cu, Ni and Zn) in Raphanus sativus L. and Spinacia oleracea L. plants irrigated with industrial effluent. *Journal of Environmental Biology, 37*(2), 381-384.

Qadir, A., Malik, R. N., & Husain, S. Z. (2008). Spatio-temporal variations in water quality of Nullah Aik-tributary of the river Chenab, Pakistan. *Environmental Monitoring and Assessment, 140*(1-3), 43-59. http://dx.doi.org/10.1007/s10661-007-9846-4

Schulz, K., & Huwe, B. (1999). Uncertainty and sensitivity analysis of water transport modelling in a layered soil profile using fuzzy set theory. *Journal of Hydroinformatics, 1*, 127-138.

Suthar, S., Chhimpa, V., & Singh, S. (2009). Bacterial contamination in drinking water: a case study in rural areas of northern Rajasthan, India. *Environmental monitoring and assessment, 159*(1-4), 43-50. http://dx.doi.org/10.1007/s10661-008-0611-0

WHO. (1984). *Solid Waste Management in South-East Asia*. WHO House, New Delhi, India

WHO. (1997). Nitrogen Oxides (2nd ed.), *Environmental Health Criteria,* Geneva, Switzerland.

WHO (Ed.) (2008). *Guidelines for Drinking-water Quality, Incorporating 1st and 2nd Addenda* (Vol. 1). World Health Organization Press, Geneva, Switzerland.

6

Dilution-Extrapolation Hydrometer Method for Easy Determination of API Gravity of Heavily Weathered Hydrocarbons in Petroleum Contaminated Soil

Carlos M. Morales-Bautista[1], Randy H. Adams[1], Francisco Guzmán-Osorio[1] & Deysi Marín-García[1]

[1] División Académica de Ciencias Biológicas (DACBiol.), Universidad Juárez Autónoma de Tabasco, Villahermosa, Tabasco, Mexico

Correspondence: Randy H. Adams, División Académica de Ciencias Biológicas, Universidad Juárez Autónoma de Tabasco (UJAT), Km 0.5 Carr. VHSA-Cárdenas, Villahermosa, Tabasco, CP 86102, Mexico. E-mail: drrandocan@hotmail.com

Abstract

When crude petroleum is spilled onto soil, the oil's properties have a large influence on the toxicity to soil organisms, the biodegradability of the oil, and potential for long term fertility problems in the soil. Furthermore, these properties of environmental concern are related to the crude's density, commonly measured as API gravity. Currently, methods do not exist to determine the °API of crude oil in contaminated soil. In this study a novel method is presented for the determination of API gravity in small volumes (< 10 ml) of heavy and extra-heavy petroleum from contaminated soil. Is uses an economical and readily available solvent (diesel + automotive lubricating oil) in a procedure based on the conventional hydrometer method, plus dilution-extrapolation techniques. It was validated with crude petroleum in the 27.1-15.0 °API range, obtaining an excellent correlation with the conventional method (R = 0.996) and an error of less than 0.4% based on specific gravity. Potential applications of this method are discussed for petroleum contaminated soil.

Keywords: analytical, °API, biodegradation, contamination, soil fertility, specific gravity

1. Introduction

In several parts of the world, ecosystem pollution has been associated with crude oil spills, usually caused by pipeline breakage, or improper extraction and transport processes (Arife et al., 2005; Bakhtiari et al., 2009; Osuji & Ezebuiro, 2006; Infante, 2001; Sakari et al., 2008; Sakari et al., 2010; Zakaria et al., 2002). Some physical and chemical properties of crude oil can cause deleterious effects in living beings (Adams et al., 2006; Eisman et al., 1991), and lower soil fertility (Adams et al., 2008; Litvina et al., 2003; Zalik et al., 2010). In the environmental area, the concentration of petroleum hydrocarbons in contaminated soil is usually determined using gravimetric extraction methods. However, these methods generally are not used to determine the *properties* of the oil in the soil.

None-the-less, petroleum engineers have known for decades that many properties of crude petroleum can be correlated to a simple parameter - the oil's density, commonly measured as API gravity. Also called API density or API degrees (°API), it is directly related to the specific gravity (density) of the oil. This parameter has been used to estimate valuable properties of this natural resource such as the quantity of easily recovered gasolines and middle distillates during refinery processes, residual content, sulphur, asphaltenes, and viscosity (Pemex Refinación, 2000; Udoetok & Osuji, 2008). Likewise, on a regional basis, the fractional content of aliphatic, aromatic, and polar compounds have been related to API density.

Some of these properties, such as asphaltene and polar content are also related to environmentally important properties of crude oil, such as the biodegradability of petroleum in soil (McMillen et al., 2002), and negative impacts to soil fertility due to changes in important physical-chemical properties of the soil (Litvina et al., 2003; Adams et al., 2008). In addition, the toxicity of petroleum has been related to low molecular weight hydrocarbons, more common in low density (high °API) petroleum (Eisman et al., 1991; Edwards et al., 1995). Thus, it can be very useful to know the °API of the oil in the soil at contaminated sites. For sites which have been recently contaminated, and in which the °API of the spilled oil is known, this information can be used to evaluate potential biodegradation, toxicity, and possible long term fertility problems in the soil. However, in older spills, where the oil has been transformed due to biodegradation and weathering, it would be very helpful to determine the API gravity of the *existing* oil in the soil (which will be different from that of the spilled oil).

Several methods can be used to determine API gravity (ASTM, 2002, 2006, 2008; Pemex Refinación, 2000). The ASTM D6822-02 thermo-hydrometer method (revised 2008) is among the most frequently used for the precise determination of API gravity in crude petroleum and petroleum products. The density is corrected to a reference temperature (60 °F = 15.6 °C) using standardized tables. This method is practical when the oil is fluid at room temperature and when relatively large volumes (~0.5 L) are available. For extra-heavy crude oil and heavy fractions of refined oil (which are not fluid at room temperature), API gravity is usually determined using specialized equipment – a Gay-Lussac pycnometer (ASTM method D369-84, revised 2002). In this method, a small amount (~10-25 ml) of oil is heated with constant mixing until fluid, and then poured into the pycnometer. As it cools, the volume is recorded at different temperatures. The density vs. temperature curve is plotted, and by applying the law of constant mass, and extrapolating to the reference temperature (60 °F), the density (°API) is thus determined. This method encounters some difficulties when very heavy, viscous hydrocarbons are used, in that it is necessary to increase the temperature to high levels (> 150 °C) and some of the hydrocarbons in the mixture may be evaporated or charred in the process. Also, the measurement of small changes in volume in very viscous mixtures is problematical and frequently imprecise.

However, these methods, by either hydrometer or pycnometer, are not practical for the determination of °API in residual concentrations in contaminated soil. In such conditions, the hydrocarbons are typically very weathered, dense, and found in relatively small concentrations (< 0.1-1%). Under these circumstances, it is very difficult and unpractical to try to extract sufficient oil from the soil to run a determination by the standard hydrometer method (which requires ~0.5 L of oil). The pycnometer method requires less oil, but still has problems with precise quantification due to problems previously mentioned.

In this study, a modified, more practical, method was developed and validated. This method uses a low cost, easily available solvent (diesel+lubricating oil) in a dilution-extrapolation procedure to determine °API in small volumes (< 10 ml) of heavy and extra-heavy petroleum, such as that typically found as a residual fraction in an oil spill after a prolonged time period or following remediation activities.

2. Experimental Work

2.1 Determination of API Gravity

All of the API gravity determinations were made using a modification of the ASTM D6822-02 method (revised 2008), with three repetitions (n = 3). Crude petroleum with API gravities of 15.03, 20.15, and 27.11 °API were used in this study. The medium crude (27.11 °API) was obtained from a tank storage facility in Comalcalco, Tabasco, (Mexico), and the heavier crude (15.03 °API) was obtained from an out-of-service sulphur well in the Texistepec Mining Unit (Unidad Minera Texistepec, Texistepec, Veracruz). From these two sources, the third API gravity oil (20.15 °API) was obtained by combining and mixing in roughly equal quantities.

A mixture of automotive diesel (PEMEX) and multigrade lubricating oil API SL SAE 20W-50 (Bardahl, Mexico, D.F.) was used as solvent. The de-sulphurized automotive diesel (ID No. UN1202, guide 13, 35 °API; Pemex Refinación, 2000) was weathered at ambient temperature (~30 °C) until enough of the lighter components had evaporated to have a relatively constant API gravity (~36 °API), as measured using an integrated hydrometer with thermometer (ASTM, 2008), calibrated according to an ASTM 54HL standard, with a range of 29-41 °API (Figure 1). Subsequently, this weathered diesel was mixed with the lubricant in proportions of approximately 1:1 (w/w) to prepare the solvent. Four hundred and fifty millilitres of this solvent were added to a 500ml graduated cylinder. In this same vessel, crude petroleum was slowly added and mixed using a glass rod until a final concentration of 0.5% (w/w) was obtained. Once complete homogenization was achieved, °API was determined with the hydrometer and the reading was corrected for a density a 60 °F (15.6 °C). In the same cylinder, more petroleum was slowly added and mixed in, with increments of approximately 0.5% (w/w), making readings with the hydrometer between additions. Increments up to only 2% were used for the heavy crude. For the medium crude, and crude oil mixture, increments up to 11% were used.

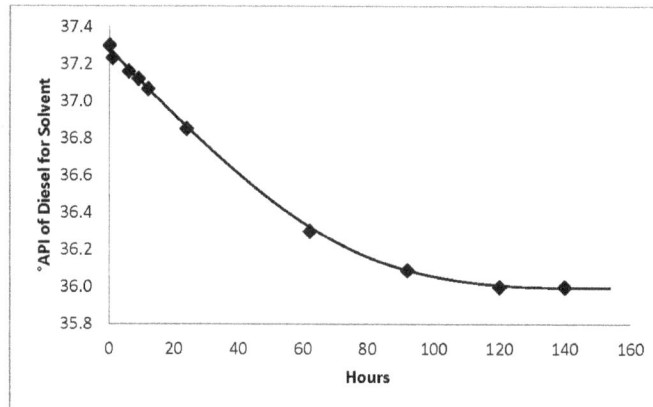

Figure 1. API gravity stabilization of the diesel used for solvent preparation

2.2 Calculations, Correlations and Statistical Comparisons

Based on the mass balance, the following equation was used:

$$(\%W_{HC} \cdot {}^{\circ}API_{HC}) + (\%W_{Solv} \cdot {}^{\circ}API_{Solv}) = (\%W_{Mix} \cdot {}^{\circ}API_{Mix})$$

Where:

$\%W_{HC}$ = weight percent of hydrocarbon in mixture,

${}^{\circ}API_{HC}$ = API gravity of hydrocarbon (in this case, crude oil),

$\%W_{Solv}$ = weight percent of solvent (in this case, diesel + lube oil),

${}^{\circ}API_{Solv}$ = API gravity of solvent,

$\%W_{Mix}$ = weight percent of mixture (crude oil + solvent),

${}^{\circ}API_{Mix}$ = API gravity of mixture (crude oil + solvent).

Considering that: the $\%W_{HC}$ represents the quantity of hydrocarbon (crude oil) mass added, that the $\%W_{Mix}$ is constant (100%), and the °API of the solvent (diesel + lube oil) is known (or measured), the above equation can be rearranged to obtain a single dependent variable:

$${}^{\circ}API_{HC} = [(100\% \cdot {}^{\circ}API_{HC}) - (\%W_{Solv} \cdot {}^{\circ}API_{Solv})]/[\%W_{HC}]$$

To obtain more precision, linear regressions were performed on the °API vs. percent weight data to obtain correlation coefficients and regression functions. These functions were then extrapolated to 100% crude oil to calculate the API gravity of the (undiluted) crude oil. Direct hydrometer readings on the three crude oils (observed °API values) were compared to the calculated values. The correlation between these values was evaluated by linear regression analysis. Additionally, the calculated vs. observed values were compared statistically using Statgraphics® ver. 5.1.

3. Results and Discussion

3.1 Comparison of Dilution-Extrapolation Hydrometer Method to Standard Hydrometer Method

The results of the determinations using the dilution-extrapolation method are shown in Figures 2-4, in which the API gravity has already been normalized to 60 °F. In these figures the relationship between percent crude petroleum ($\%W_{HC}$) in the solvent (diesel + lubricating oil) is graphed against °API (of the mix of crude oil + solvent) for the three crude petroleums used (15.03, 20.15 and 27.11 °API). Linear regressions are observed with correlations of $|R| > 0.98$, showing less uncertainty in the heavier crudes ($|R| = 0.994$) than in the medium crude ($|R| > 0.983$). This is due to the inverse relationship between API gravity and specific gravity. With the lighter petroleum, more is needed to be able to observe a notable difference on the hydrometer, thus the smaller difference in the API gravity measurement in the lighter (less dense) petroleum, and the lower precision obtained by visual measurements of smaller differences. For this reason is was necessary to add more petroleum to the solvent, up to a final concentration of about 11% (w/w), versus only about 2% for the heavy crude. However, this may be of little practical importance, at least in most older spills, due to the fact that the great majority of soils contaminated with petroleum in these environments have predominately weathered hydrocarbons, rich in heavy and extra-heavy oil (Adams & Morales, 2008; Udoetok & Osuji, 2008; Litvina et al., 2003). With this method it is possible to determine the API gravity of oil in contaminated soil using less sample, even when the concentration of hydrocarbons in the soil is relatively low (< 1%), and when the soil is contaminated with heavy and extra-heavy oils.

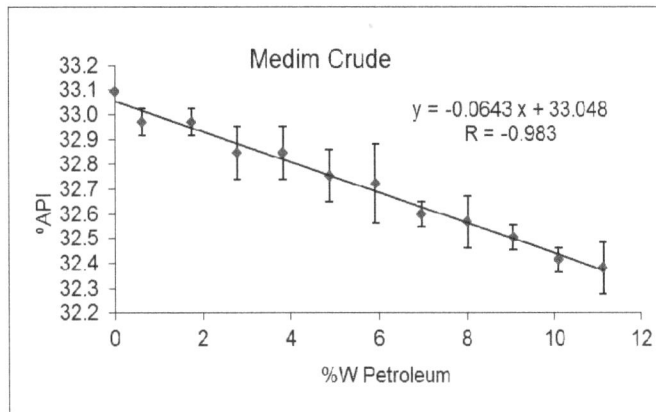

Figure 2. Relationship between °API and weight percent of medium crude petroleum
Note: Error bars represent one standard deviation.

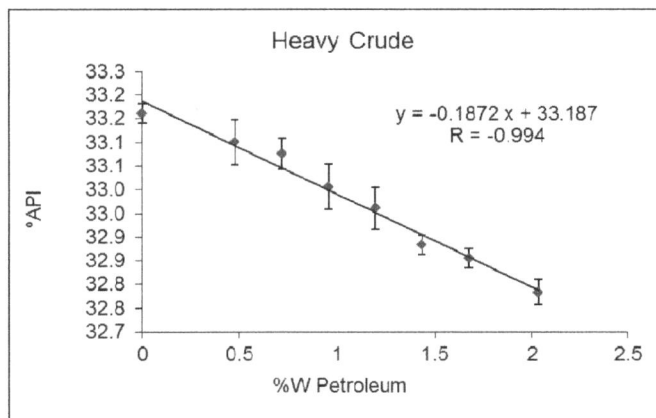

Figure 3. Relationship between °API and weight percent of heavy crude petroleum
Note: Error bars represent one standard deviation.

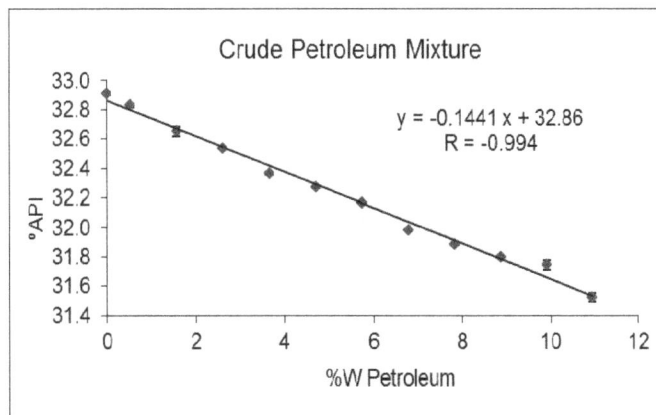

Figure 4. Relationship between °API and weight percent of crude petroleum mixture
Note: Error bars represent one standard deviation.

The correlation between the calculated values obtained by this novel method and the conventional method is shown in Figure 5. As seen in this figure, there is an excellent correspondence (R=0.9996) with the function constrained to pass through zero, and with a slope of nearly 1 (1.0002). In this figure, the error bars are so small as to be nearly imperceptible. The difference between these methods, based on specific gravity, was less than 0.4%. The statistical comparison of the direct method vs. the dilution-extrapolation method showed no significant difference for all three oils (no overlap in Standard Deviation at a 99% Confidence Limit, Figures 6-8). The slight differences that were observed, were much less in the heavy crude in comparison with the medium crude or crude oil mixture. A Kruskal-Wallis test of the data indicated a p-value of 0.0369 with no significant differences at a 99% confidence level.

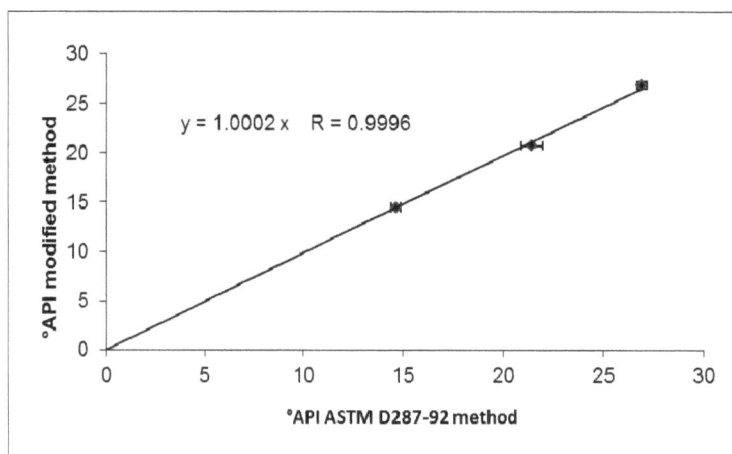

Figure 5. Correlation between dilution-extrapolation method and direct measurement with hydrometer
Note: Error bars represent one standard deviation (nearly imperceptible).

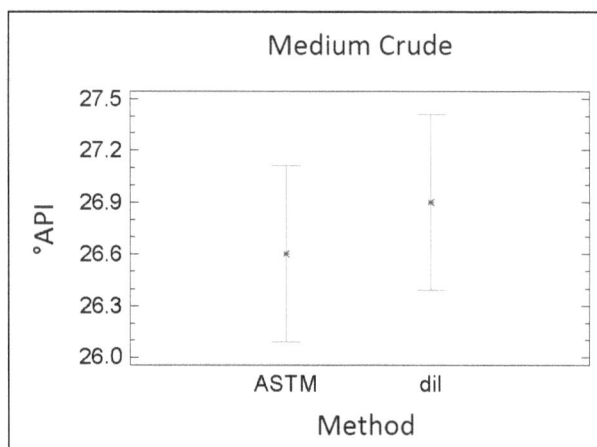

Figure 6. Statistical comparison between the ASTM method vs dilution extrapolation (medium crude petroleum)
Note: Error bars represent 99% confidence intervals (mean difference).

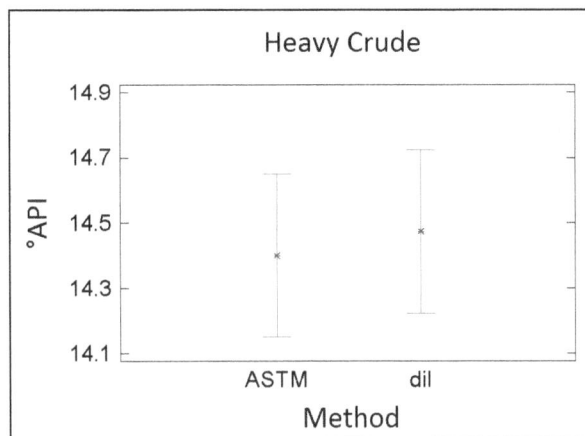

Figure 7. Statistical comparison between the ASTM method vs dilution extrapolation (heavy crude petroleum)
Note: Error bars represent 99% confidence intervals (mean difference).

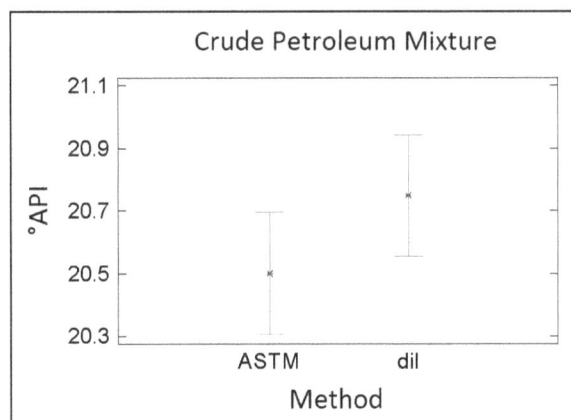

Figure 8. Statistical comparison between the ASTM method vs dilution extrapolation (crude petroleum mixture)
Note: Error bars represent 99% confidence intervals (mean difference).

3.2 Utility of Novel Method for Characterization of Petroleum Contaminated Soils

The determination of °API of weathered oil in contaminated soil can be very practical for the evaluation of many properties of environmental concern. For example, it is well known that the more viscous and weathered the oil is (lower °API), it is much more difficult to biodegrade. McMillen et al. (2002), have shown that the bioremediation endpoint is directly proportional to the °API of the oil spilled (R=0.978), and crudes with a value of about 30 °API and higher are readily biodegradable, whereas oils with °API values of 20 or less are very difficult and slow to biodegrade. Likewise, the presence of more polar functional groups in the residual oil, may enhance the formation of thin layers of oil on the surfaces of soil particles, and thereby cause water repellency (Litvina et al., 2003). This process may also lead to reduced field capacity, and compaction in the soil (Adams et al., 2008, Trujillo-Narcia et al., 2012). Such polar groups are more common in weathered (low °API) oil, which has a higher proportion of polar compounds and asphaltenes (McMillen et al., 2002), and much greater viscosity (Ancheyta et al. 2011). Thus, in contaminated soil containing a residual oil with lower °API values, there may be a much greater probability of long term fertility problems.

As a simple test of the utility of determining the °API of oil in contaminated soil for environmental purposes, we evaluated the correlation between °API and two properties of environmental importance (viscosity and S content). Using existing data, we evaluated 10 crudes from South East Asia, from oil fields in central Sumatra, the Gulf of Thailand, the Natuna Sea, and the South China Sea, with a range of 20 – 60 °API, (Chevron Corporation, 2011; Table 1 and Figure 9). In this data set, the sulphur content (and therefore the relative proportion of polar functional groups in the oil, and thus, potential for soil fertility problems) was highly correlated to °API (($|R|$=0.969, Figure 10a). Likewise, the viscosity (and therefore biodegradability), was also logarithmically related to °API with a high correlation ($|R|$=0.955, Figure 10b).

Table 1. Properties of some crude oils from SE Asia

Crude Oil	Area	°API	Viscosity (cSt at 50°C)	Wt% S
Duri	Central Sumatra	20.29	205.40	0.21
Minas	Central Sumatra	33.94	9.72	0.09
Sembilang	Natuna Sea	35.20	11.59	0.05
Pattani	Gulf of Thailand	38.11	3.78	0.06
Nanhai Light	S China Sea (Hong Kong area)	39.50	4.03	0.06
Tantawan	Gulf of Thailand	41.64	2.90	0.05
Benchamas	Gulf of Thailand	41.82	4.17	0.04
Belida	Natuna Sea	46.60	1.99	0.02
Belanak	Natuna Sea	53.17	1.40	0.01
Erawan	Gulf of Thailand	58.90	0.60	0.01

Chevron Corporation (2011). Chevron Crude Oil Marketing/Far Eastern Crudes – Whole Crude Properties. Retrieved from http://crudemarketing.chevron.com/crude/far_eastern

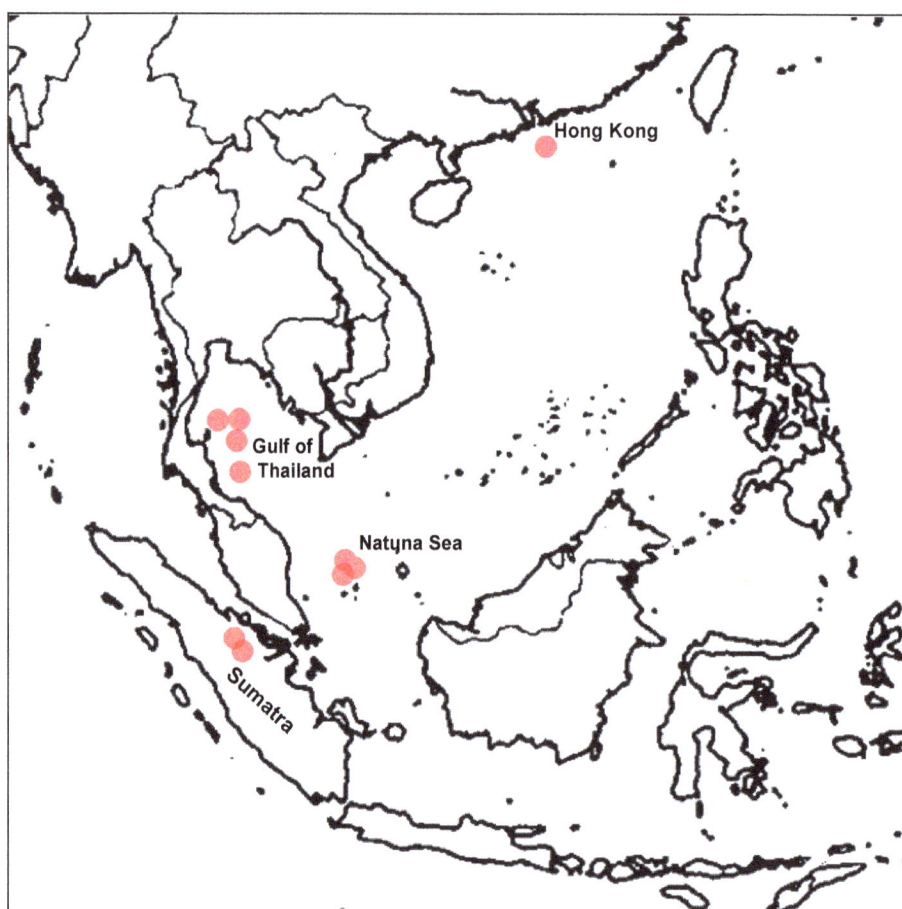

Figure 9. Approximate locations of SE Asian oil fields used in °API utility evaluation

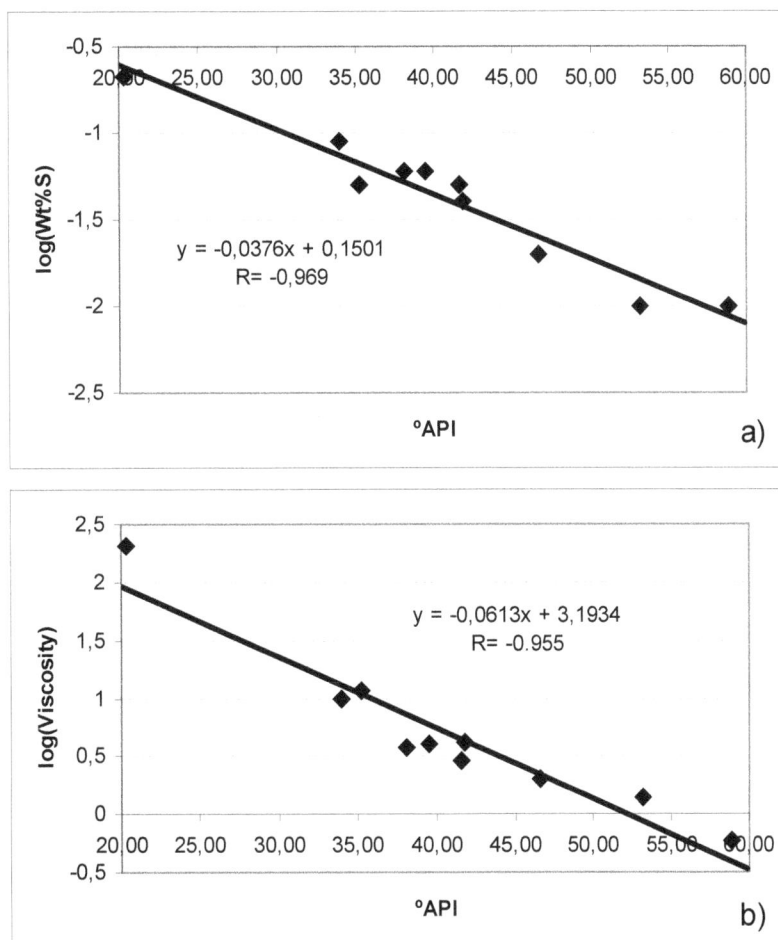

Figure 10. Correlation between °API and sulphur content (a)), and viscosity (b)) of crude oils from SE Asia

Note: Sulphur content as weight percent; viscosity at 50°C in cSt. Correlations are logarithmic.

These correlations indicate that the biodegradability and potential long term fertility problems of soil contaminated with crude oil could be related to the °API of the oil in the soil. In south and SE Asia, this might occur in on-shore oil fields, such as those in central Sumatra, or upper Assam state (India). In addition, these correlations could be used to evaluate the biodegradability of sludge in on-shore processing facilities, such as those near Johor Bahru (Malaysia) or in Brunei.

Another possible use of this method would be to determine the potential for biodegradation (or long term fertility problems) of the oily viscous fluids (OVF) produced in the steam flooding operations for the extraction of heavy oil, such as those in Alberta (Canada), Venezuela, or Sumatra (Indonesia), (Mai & Kantzas, 2009; Souraki et al., 2012, 2013; Arfie et al., 2005). These residues are typically composed of a mixture of heavy oil, sand, and water emulsions. The ecological treatment of these kinds of wastes is very problematical in production and processing areas (Arfie et al., 2005). Previous knowledge about the kind of petroleum left in the residue (as opposed to the oil in the reservoir) would be useful to develop treatment strategies that have better logistics or costs benefits.

4. Conclusion

The novel method developed in this study provides a simple but reliable analytical tool for the environmental characterization of crude oil in soil, something that was not possible with previously existing methods. Comparing this novel method for °API determination by dilution-extrapolation with the conventional method with hydrometer, there was an excellent correlation (R=0.9996) and a very low error (< 0.4%). In quantitative terms, these methods are practically equivalent. However, with the dilution-extrapolation method it is feasible to use very small volumes of oil (< 10 ml) facilitating its application to soils with low hydrocarbon concentrations (0.1-1.0%). Likewise, it is a practical method for the determination of API gravity in heavy and extra-heavy oils, commonly found in soils contaminated by residual hydrocarbons. Thus, the °API of oil in contaminated soil can be easily determined by recovering small amounts of petroleum from soil using solvent extraction procedures (EPA, 1996), and then applying the dilution-extrapolation method on the recovered oil. For these reasons, it is recommended as a viable alternative for the determination of API gravity in hydrocarbon contaminated soils, and a useful tool for the characterization of contaminated sites, with relevance to the biodegradability of the oil in the soil, and the potential to cause long term fertility problems.

Acknowledgements

We would like to acknowledge the Consejo Nacional de Ciencia y Tecnología (National Council of Science and Technology, Mexico) and the Government of Tabasco State for financing this research project, Fomix 2009-04, TAB-2009-C18-121493.

References

Adams, R., Álvarez-Ovando, A. L., Escalante-Espinosa, E., & Gutiérrez-Rojas, M. (2006). Dose-Response relationship of organisms in soil with heavily weathered hydrocarbons and changes in fertility parameters. *13th International Petroleum Environmental Conference* (pp. 1-7), San Antonio, Texas. Retrieved from http://ipec.utulsa.edu/Conf2006/Papers/Adams_37.pdf

Adams, R., & Morales-García, F. (2008). Concentración residual de hidrocarburos en suelo del trópico: Parte I. - Consideraciones para la salud publica y protección al ganado. *Interciencia, 33*(7), 476-482. Retrieved from http://www.interciencia.org/v33_07/476.pdf

Adams, R., Zavala-Cruz, J., & Morales-García, F. (2008). Concentración residual de hidrocarburos en suelo del trópico: Parte II. - Afectación a la fertilidad y su recuperación. *Interciencia, 33*(7), 483-489. Retrieved from http://www.interciencia.org/v33_07/483.pdf

American Society for Testing and Materials International. (2002). *ASTM D369-84(2002) Standard test method for specific gravity of creosote fractions and residue.* Pennsylvania: West Conshohocken.

American Society for Testing and Materials International. (2006). *ASTM D287-92(2006) Standard test method for API gravity of crude petroleum and petroleum products (hydrometer method).* West Conshohocken, Pennsylvania.

American Society for Testing and Materials International. (2008). *ASTM D6822-02(2008) Standard test method for density, relative density, and API gravity of crude petroleum and liquid petroleum products by thermohydrometer method.* Pennsylvania: West Conshohocken.

Ancheyta, J., Centeno, G., Sánchez-Reyna, G., & Nájera, A. (2011). Cálculo de viscosidad de petróleo mediante real de mezclado con parámetros de interacción binaria. *LI Convención Nacional – Instituto Mexicano de Ingenieros Químicos A.C.,* 20-21 Octubre; Puebla, México. Sesión JM-2-3. Retrieved from http://imiq.com.mx/convencion/web/SESSIONES/JM-3-2.pdf

Arfie, M., Marika, E., Purbodiningrat, E. S., & Wooda, H. A. (2005). Implementation of slurry fracture injection technology for E&P wastes at Duri oilfield. *SPE Asia Pacific Health, Safety and Environment Conference and Exhibition,* Kuala Lumpur, Malaysia, 19-20 September 2005. Paper: SPE 96543-PP. Retrived from http://www.terralog.com/article/SPE-96543-PP.pdf#zoom=130%25

Bakhtiari, A. M., Zakaria, M. P., Yaziz, M. I., Lajis, M. N. Hj, & Bi, X. (2009). Polycyclic aromatic hydrocarbons and *n*-alkanes in suspended particulate matter and sediments from the Langat River, Penisular Malaysia. *EnvironmentAsia, 2,* 1-10.

Chevron Corporation. (2011). *Chevron Crude Oil Marketing/Far Eastern Crudes – Whole Crude Properties.* Retrieved from http://crudemarketing.chevron.com/crude/far_eastern

Edwards, D. A., Andriot, M. D., Amoruso, M. A., Tummey, A. C., & Bevan, C. J. (1995*). Development of fraction specific reference doses (RfDs) and refernce concentrations (RfCs) for total petroleum hydrocarbons (TPH), vol. 4.* The Association of Environmental Health and Sciences. Retrieved from http://library.wur.nl/WebQuery/clc/1658752

Eisman, M. P., Landon-Arnold, S., & Swindoll, C. M. (1991). Determination of petroleum hydrocarbon toxicity with Microtox®. *Bulletin of Environmental Contamination and Toxicology, 47,* 811-816. http://dx.doi.org/10.1007/BF01689508

Infante, C. (2001). Biorestauración de áreas impactadas por crudo por medio de Intebios® y Biorize®. *Interciencia, 26*(10), 504-507. Retrieved from http://www.interciencia.org/v26_10/index.html

Litvina, M., Todoruk, T. R., & Langford, C. H. (2003). Composition and structure of agents responsible for development of water repellency in soils following oil contamination. *Environmental Science and Technology, 37*, 2883-2888. http://dx.doi.org/10.1021/es026296l

Mai, A., & Kantzas, A. (2009). Heavy oil waterflooding: effects of flow rate and oil viscosity. *J. Can. Pet. Technol, 48*(3), 42-51.

McMillen, S., Smart, R., & Bernier, R. (2002). Biotreating E & P wastes: lessons learned from 1992-2002. *9th International Petroleum Environmental Conference*, Albuquerque, New Mexico; pp. 1-14. Retrieved from http://ipec.utulsa.edu/Conf2002/mcmillen_smart_bernier_122.pdf

Osuji, L. C., & Ezebuiro, P. E. (2006). Hydrocarbon contamination of a typical mangrove floor in Niger Delta, Nigeria. *International Journal of Environmental Science and Technology, 3*(3), 313-320. Retrieved from http://www.ijest.org/?_action=aricleInfo&article=133

Pemex Refinación. (2010). *Diccionario de términos de PEMEX Refinación.* México D. F. Retrieved from http://www.itek.com.mx/INDUSTRIA/DICCIONARIO%20PEMEX.pdf

Safari, M., Zakaria, M. P., Mohamed, Che Abd R., Lajis, N. Hj., Chandru, K, Bahry, P. S., ... Anita, S. (2010). The history of petroleum pollution in Malaysia; urgent need for integrated prevention approach. *EnvironmentAsia, 3*(special issue), 131-142.

Sakari, M., Zakaria, M. P., Lajis, N., Mohamed, C. A. R., Shahpoury, P., Anita, S., & Chandru, K. (2008). Characterization, distribution, sources and origins of aliphatic hydrocarbons from surface sediments of Prai Straits, Penang, Malaysia: A widespread anthropogenic input. *EnvironmentAsia, 1*(2), 1-14.

Souraki, Y., Ashrafi, M., Karimaiel, H., & Torsaeterl, O. (2012). Experimental analyses of Athabasca bitumen properties and field scale numerical simulation study of effective parameters on SAGD Performance. *Energy and Environment Research, 2*(1), 140-156. http://dx.doi.org/ 10.5539/eer.v2n1p140

Souraki, Y., Ashrafi, M., & Torsaeterl, O. (2013). A comparative field-scale simulation study on feasibility of SAGD and ES-SAGD processes in naturally fractured bitumen reservoirs. *Energy and Environment Research, 3*(1), 49-62. http://dx.doi.org/10.5539/eer.v3n1p49

Trujillo-Narcía, A., Rivera-Cruz, M. d C., Lagunes-Espinoza, L. d C., Palma-López, D. J., Soto-Sánchez, S., & Ramírez-Valverde, G. (2012). Efecto de la restauración de un fluvisol contaminado con petróleo crudo. *Revista Internacional de Contaminación Ambiental, 28*(4), 361-374. Retrieved from http://www.revistas.unam.mx/index.php/rica/aricle/view/25347

Udoetok, I. A., & Osuji, L. C. (2008). Gas chromatographic fingerprinting of crude oil from Idu-Ekpeye oil spillage site in Niger-delta, Nigeria. *Environmental Monitoring and Assessment, 141*, 359-364. http://dx.doi.org/10.1007/ s10661-007-9902-0

United Estates Environmental Protection Agency. (1996). *Method 3540C. Soxhlet extraction.* SW-846 Manual. Washington, DC: Government Printing Office. Retrieved from http://www.epa.gov/osw/hazard/testmethods/sw846/pdfs/3540c.pdf

Zakaria, M. P., Takada, H., Kumata, H., Nakada, N., Ohno, K., & Mihoko, Y. (2002). Distribution of Polycyclic Aromatic Hydrocarbons (PAHs) in rivers and estuaries in Malaysia: widespread input of petrogenic hydrocarbons. *Environmental Science and Technology, 36*, 1907-18. http://dx.doi.org/10.1021/es011278+

Zalik, A. (2010). Volatilidad y mediación en diferentes campos petroleros: las arenas bituminosas y la delta de Niger como lugares de controversia. *Revista Umbrales Ciencias Sociales, 20*, 307-334. Retrieved from http://www.revistasbolivianas.org.bo/pdf/umbr/n20/n20n20a11.pdf

7

Health Effects of Coal: A Long-Run Relationship Assessment of Coal Production and Respiratory Health in Kazakhstan

Almaz Akhmetov[1,2]

[1] ENCA Management, Yessik, Kazakhstan

[2] Orizon Consulting Services, McLean, VA, USA

Correspondence: Almaz Akhmetov, ENCA Management, Pobedy Str., 7, Yessik, 040400, Kazakhstan. E-mail: al_akhmetov@yahoo.co.uk

Abstract

Respiratory diseases, like asthma and other chronic obstructive pulmonary disease, claim over 50 thousand lives annually in Kazakhstan according to national statistics (The Agency of Statistics of the Republic of Kazakhstan, 2011). This study applies econometric methods to examine the relationship between the coal industry and the respiratory health in Kazakhstan during the country's independency period using annual national data. The study investigates long-term equilibrium and short-term dynamics of coal production and respiratory diseases in Kazakhstan by applying the Vector Error Correction Model (VECM). The empirical results show that the respiratory diseases appear to be elastic relative to the coal production, and the strong long-run and short-run Granger causality running from coal production to respiratory diseases. The presence of causal relationship could be useful to define effective policies to reduce the health effects of coal industry in Kazakhstan.

Keywords: Kazakhstan, coal industry, respiratory health, granger causality, air pollution

1. Introduction

About 5 million people in Kazakhstan live in an area with polluted air, and the cost of health damage from the air pollution reaches 70.8 USD per person, or 76.2 USD per ton of air pollutant (Zubov, 2007). The major source of the air pollution in Kazakhstan is the energy sector, which is primarily fuelled by coal. Coal is the most abundant fossil fuel on Earth, and Kazakhstan's reserves make up almost 4% of the world`s total coal reserves (BP, 2011). Coal is being actively replaced by cleaner sources of energy worldwide and Kazakhstan is among countries implementing a low-carbon development strategy and designing a law aimed at promoting renewable energy. Despite this fact, Kazakhstan is most likely to maintain the current status quo with regards to the coal consumption due to significantly low cost of extraction and transportation of the coal which makes it a dominant source of energy in Kazakhstan.

Coal accounts for about 44% of total primary energy demand and electricity production as well as 80% of the total power production in Kazakhstan (IEA, 2010). With exception of western Kazakhstan, where oil and gas are the main products used for the power production, in all other parts of the country coal is the primary source for the power generation and space heating. Approximately 60% of the coal in Kazakhstan is used as the steam coal at the power plants, 14% of coal is used to heat space and water by residential sector and 26% of coal mined in Kazakhstan is exported to other countries, primarily to Russia and Kyrgyz Republic (Concept of Coal Industry Development of Kazakhstan until 2020, 2008). The electricity and heat are co-generated through the use of the steam coal at large scale State District Power Stations (SDPS) and Combined Heat and Power (CHP) plants. The largest coal producers are the structural units of large power and metallurgical companies. The district boiler stations burn coal to produce heat for the centralized residential heating system.

The general characteristics of Kazakhstani coal are high calorific properties and low sulfur content. However, the high ash content of the coal not only lowers its competitiveness in the world market, but also affects the stability of the combustion process and the reliability of furnaces at the power plants. This results in increased emissions of carbon monoxide, nitrogen oxides and sulfur oxides (Bukhman, 2003). In the life cycle of coal consumption, pre-combustion emissions of hazardous substance to air contribute only 0.2% of the total pollution, while the rest of the air pollution is created during the combustion process (Akhmetov et al., 2012). However, soil and water pollution is not accounted for in the existing studies in Kazakhstan.

Due to the unequal distribution of the reserves of fossil fuel in Kazakhstan, there are areas in the country where coal is the only source of energy, particularly in the heating season. This fact also produces large seasonal cost variations in such areas. Individual residences, where no centralized heating system is available, burn coal in self-made coal stove to heat their homes. Coal combustion in the residential sector for space and water heating is not inventoried as point sources. However, this process is the main source of the indoor air pollution in coal-dependent regions of Kazakhstan. Most domestic coal stoves are of poor quality as shown in Figure 1.

Figure 1. Typical home-made coal stoves used for space heating in rural areas of Kazakhstan without centralized heating system

Visible cracks and dirt deposits on the stoves may indicate the presence of indoor air pollution. Furthermore, the low temperature of the combustion and short chimneys may indicate that air pollution disperses downward and increases near the ground. Hence, the pollution is localized and causes a degradation of the regional environment.

Respiratory diseases are the most commonly diagnosed diseases in Kazakhstan, and account for 42% of the total registered diseases (The Agency of Statistics of the Republic of Kazakhstan, 2011). The statistics indicate that respiratory disease causes substantial burden on the economy and health of the country. In 2004, the disability-adjusted life year (DALY), an overall disease burden measure, due to non-communicable respiratory diseases was the 9th biggest in the world, or 1.6 times bigger than world average (WHO, 2009). Although there are different causes of respiratory diseases, the coal industry is very often the main cause in regions, where coal is the main fuel (Lockwood, 2012, pp. 111-127).

The existing research on air pollution and health deterioration are mainly regional and city level studies (Bowen et al., 1995; Koop et al., 2010; Mennis, 2005; Sheppard, et al. 1999; Sobral, 1989). The studies indicate a strong relationship between air pollution and induced health problems in the short- and long-term. The existing studies (Gelobter, 1992; Gianessi et al., 1979) also support the evidence of the relationship between health conditions and air quality. Furthermore, studies on indoor air pollution impact on health in developing countries (Dasgupta et al., 2006; Smith & Mehta, 2003) also prove the connection between the variables. Most of the regional and national studies have identified that poor people, ethnic minorities, children and elderly people are most vulnerable to air pollution impacts. Ren & Tong (2008) made a thorough overview of recent epidemiology research developments and methodological issues on health effects of ambient air pollution.

Health impacts of coal industry from various locations worldwide are comprehensively described by Finkelman et al. (2002). The examples provided in the study suggest that coal-related health problems are becoming serious issue in emerging and developing countries, where cheap coal is the main fuel for the economies. Generally, the studies about the coal industry impact on health and well-being of people could be divided by life cycle stages of coal: mining and storage (Ghose & Majee, 2000; Hendryx, 2009; Morrice & Colagiuri, 2013), combustion at the power plants (Penney et al., 2009; Riekert & Koch, 2011) and full life cycle (Castelden et al., 2011; Epstein et al., 2011).

The literature on health effects of the coal industry in Kazakhstan primarily describes the health status of the coal miners (Terekhin & Pichkhadze, 1991). Dahl & Kuralbayeva (2001) indicated that the coal production and use are the main causes for the environmental degradation in the industrial regions of Kazakhstan.

Kenessariyev et al. (2013) estimated the mortality attributed to air pollution caused by total suspended particles in 11 cities across Kazakhstan. The study utilizes a log-linear concentration response function to estimate air pollution attributed mortality with other mortality causes in the country. The study revealed that the premature mortality caused by air pollution in Kazakhstan is significantly higher than in Russia and Ukraine. It was suggested that coal consumption was the main cause of such result. Furthermore, the results indicated that Almaty had the highest number of deaths attributed to air pollution in Kazakhstan. Despite the limitations of the study (such as significant uncertainties, use of total mortality without looking at the causes and not including other pollution compounds) it laid a basis for a scientific foundation for further studies on air pollution effects in Kazakhstan. To the best knowledge of the author, there is no research about the possible impact of the coal production and use on the respiratory health in Kazakhstan.

The empirical analysis presented here helps to define which factor is the main cause of growing number of respiratory diseases in Kazakhstan. The variables tested are production of commonly produced fossil fuel and the number of automobiles in Kazakhstan. In this study, fossil fuel production data is used instead of consumption data as coal industry has negative impact on respiratory health throughout the entire life cycle.

2. Method

The empirical assessment is based on testing long-term relationship between registered respiratory disease instances and production of different fossil fuels (coal, oil and natural gas) and number of automobiles in Kazakhstan, while utilization of the Vector Error Correction Model (VECM) helps to define the short-term dynamics of the variables.

2.1 Data

This study collects annual data on total respiratory diseases, coal production, oil production, natural gas production and number of automobiles for the period between 1990 and 2009. The data source for respiratory disease is the medical statistics database developed by MedInform Ltd. (MedInform Ltd., 2014). Information on oil and natural gas production is derived from the BP Statistical Review of World Energy 2011 (BP, 2011), while data on coal production is taken from the Agency of Statistics of the Republic of Kazakhstan (The Agency of Statistics of the Republic of Kazakhstan, 2011). Summary statistics of variables used in the study are given in Table 1.

Table 1. Summary statistics of variables, 1990-2009

Variable	Mean	SD	CV (%)
Total respiratory diseases	4,220,490.95	639,033.59	15.14
Coal production (thousand tons)	92,918.61	21,105.11	22.71
Oil production (million tons)	41.55	19.90	47.91
Natural gas production (billion m^3)	9.51	4.64	48.81
Number of automobiles (thousand)	1689.39	585.80	34.68

Note. SD is standard deviation; CV is the coefficient of variation.

The trends of time series shown in Figure 2 and indicate steady decline in all series from the beginning in 1990 to almost the end of 1990s and start to increase in 2000, with natural gas production exhibiting the most related variation and respiratory disease exhibiting the least related variation as displayed in Table 1. Oil and natural gas production for export is the main source of the economic growth in Kazakhstan and has significantly increased since 1990, while coal production volumes are still lower than in the pre-independence period (1990-1991). About 50% of natural gas and almost 90% of all oil produced in Kazakhstan go for export (BP, 2011), while coal is predominantly for domestic use. Growing population wealth resulted in a significant increase of the number of cars in the 21st century, and number of total diagnosed respiratory diseases has been steadily increasing too since 2000.

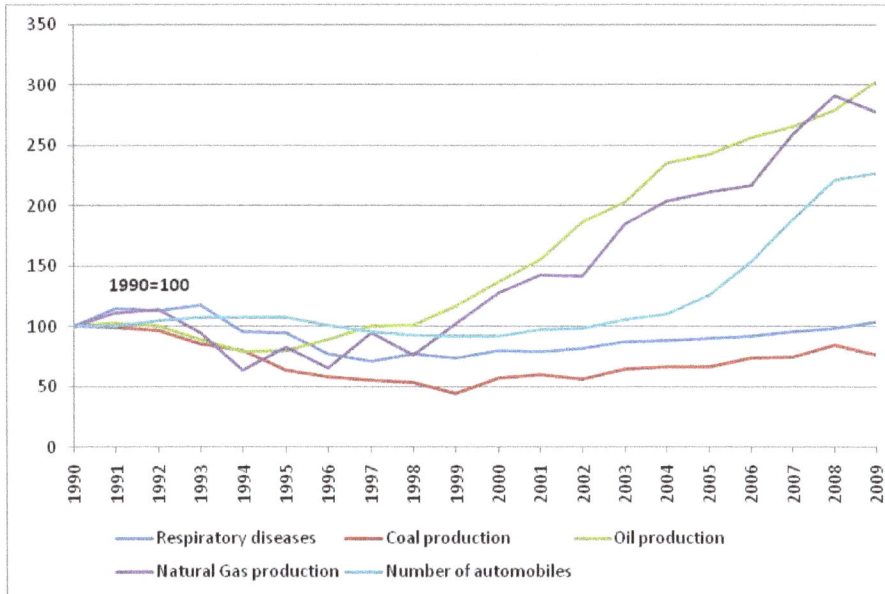

Figure 2. The series plots of total respiratory diseases, fossil fuels production (by type) and number of cars in Kazakhstan, 1990-2009

2.2 Methodology

For testing purposes, all data was converted into natural logarithms and the results can be interpreted in growth terms. The analysis is performed using ordinary regression analysis as follows:

$$LRD_t = \beta_1 + \beta_2 LCP_t + u_t \tag{1}$$

$$LRD_t = \beta_3 + \beta_4 LOP_t + u_t \tag{2}$$

$$LRD_t = \beta_5 + \beta_6 LGP_t + u_t \tag{3}$$

$$LRD_t = \beta_7 + \beta_8 LA_t + u_t \tag{4}$$

where LRD_t, LCP_t, LOP_t, LGP_t and LA_t represent natural logarithms of total respiratory diseases, coal production, oil production, natural gas production and number of automobiles for year t respectively. This step allows to select the independent variable that creates long-term equilibrium with dependent variable LRD. In other words, which of the causes of air pollution has the biggest impact on respiratory diseases in the long-term?

In order to test short-run dynamics of the variables, the VECM is utilized. The analysis is performed in four steps:

a) Verification of the order of the integration of the variables as the co-integration test is only valid for variables of the same order of integration. The Augmented-Dickey-Fuller (ADF) (Said & Dickey, 1984) and Kwiatkowski-Phillips-Schmidt-Shin (KPSS) (Kwiatkowski et al., 1992) tests are used for the purpose. While ADF test is a test for a unit root, KPSS test is designed on the basis of the null hypothesis that a series is stationary.

b) When all of the series are integrated at the same order, the Johansen maximum likelihood method (Johansen, 1991) is used to test the co-integration between the variables. The co-integration of variables indicates the presence of long-run equilibrium relationship between the variables.

c) The VECM is used to correct disequilibrium in the co-integrated relationship by means of error-correction term (ECT), as well as test for presence and direction of long- and short-run Granger causality among co-integrated variables. The VECM for Eq. (1) is specified as follows:

$$\Delta LRD_t = \gamma_{10} + \sum_{i=1}^{m_1} \gamma_{11i}\Delta LRD_{t-i} + \sum_{i=1}^{n_1} \gamma_{12i}\Delta LCP_{t-i} + \delta_1 ECT_{t-1} + \mu_{1t} \tag{5}$$

$$\Delta LCP_t = \gamma_{20} + \sum_{i=1}^{m_2} \gamma_{21i}\Delta LRD_{t-i} + \sum_{i=1}^{n_2} \gamma_{22i}\Delta LCP_{t-i} + \delta_2 ECT_{t-1} + \mu_{2t} \tag{6}$$

where

$$ECT_{t-1} = LRD_{t-1} - \alpha_0 - \alpha_1 LCP_{t-1} \tag{7}$$

The sign Δ is the first-difference operator; the optimal lag lengths m_i and n_i are determined using Akaike information criterion (AIC); and coefficients δ_1 and δ_2 measure the speed of return to equilibrium of the variables LRD and LCP respectively.

d) In the last step, the quality and robustness of the VECM model presented in Eqs. 5-7 is assessed. Known problem with the AIC-based VECM model is the possible model mis-specification caused by unstable parameters (Narayan & Smith, 2005; Hsiao-Tien et al., 2011). Hence, the parameters consistency needs to be addressed by using the cumulative sum of recursive residuals (CUSUM) and the CUSUM of square (CUSUMSQ) tests (Brown et al., 1975).

3. Results

The results of regression analysis indicate that coal production has a much greater impact on respiratory diseases than other variables. Coal production contributed to the increase of the respiratory diseases to 81.8%, followed by road transport which contributed to only 14.2% of the respiratory diseases as shown in Table 2.

Table 2. Coefficients of Eq. (1)-(4)

	Independent variables						
	LCP	LOP	LGP	LA	Intercept	R^2	95% CI
Eq. (1)	0.603*** (8.980)				8.365*** (10.916)	0.818	0.462-0.744
Eq. (2)		0.007 (0.089)			15.221*** (55.916)	0.001	-0.150-0.163
Eq. (3)			0.059 (0.823)		15.117*** (95.425)	0.036	-0.09-0.211
Eq. (4)				0.198 (1.724)	13.784*** (16.260)	0.142	-0.043-0.439

Note. Numbers in parenthesis indicate t-statistics. *** indicate a 1% level of significance. CI – confidence interval.

The insignificance of the oil and natural gas impact on respiratory disease could be explained by the fact that the fossil fuel is the primary source of the export and coal is the main domestic fuel of the economy. Furthermore, it is likely that the variable LA is more significant for urban areas.

The time series properties of LRD and LCP are checked through both ADF and KPSS unit root tests. The results of both tests indicate that both series appear to contain unit root in their levels but stationary in their first difference, indicating that they are integrated at order one i.e. $I(1)$ as displayed in Table 3.

Table 3. Results of unit root tests

	ADF test		KPSS test	
Variable	Level	1st difference	Level	1st difference
LRD	0.745	-2.577**	0.569***	0.070
LCP	-0.570	-3.793***	0.322***	0.072

Note. ** and *** indicate that the null hypothesis is rejected at 5% and 1% level respectively. The optimal lag lengths are selected using AIC.

The results of Johansen's co-integration test the presence of co-integration between the variables as seen in Table 4. The results indicate that there is a long-run equilibrium relationship between respiratory diseases and coal production, and the normalized co-integrating vector with respect to LRD is (1, 0.603) as in Table 2. This implies that 1% increase in coal production results in 0.603% increase in respiratory diseases. Hence, the respiratory diseases appear to be coal production elastic. The evidence of co-integration also indicates that the estimated model in Eq. 1 does not lead to spurious regression results, and the estimated parameters are super-consistent.

Table 4. Results of Johansen's co-integration test

Variables: LRD and LCP

Eigenvalue	Trace Statistic	5% critical value	Max. Eigen Statistic	5% critical value	Number of co-integration
0.946	47.528**	15.495	43.777**	14.264	None
0.221	3.751	3.841	3.751	3.842	At most 1

Note. The optimal lag lengths are selected using AIC. ** indicates the rejection of a null hypothesis at 5% level of significance.

The existence of co-integration between the variables indicates that Granger causality exists at least in one direction (Engle & Granger, 1987). However, it does not indicate the direction of causality. The VECM helps to define the direction of causal relationship. The short-run F-statistics, long-run t-statistics and joint F-statistics for Eqs. 5-7 are reported in Table 5.

Table 5. Results of causality tests

Dependent variables	Source of causation				
	Short-run F-statistics		Long-run t-statistics	Joint short-run and long-run F-statistics	
	ΔLRD	ΔLCP	ECT	ΔLRD/ECT	ΔLCP/ECT
ΔLRD		4.755**	-3.918**		9.471**
ΔLCP	4.211*		-1.778	4.018*	

Note. The optimal lag length is four. * and ** indicate 10% and 5% level of significance respectively.

The short-run dynamics suggest bidirectional causality from coal production to respiratory diseases and vice-versa. However, the significance of the causal relationship from coal production to respiratory diseases seems to be stronger. The estimated coefficients for ECT indicate the presence of unidirectional causality from coal production to respiratory diseases in the long-run. The joint statistics indicate Granger endogeneity and presence of strong causality from coal production to respiratory diseases. Significantly higher lag length (4) defined by AIC (the most common lag length for annual time series is one) could be partially explained by the nature of the variables as the development of respiratory diseases requires exposure period to air pollution.

CUSUM and CUSUMSQ statistics plots for the variables LRD and LCP are presented in Figures 3 and 4. As it can be seen, both statistics are well within the critical bounds of 5% significance. This implies that the estimated coefficients of the VECM model are stable for duration of the estimation period. Hence, the results of the Granger causality tests based on the VECM model can be used for policy decision-making (Hsiao-Tien et al., 2011).

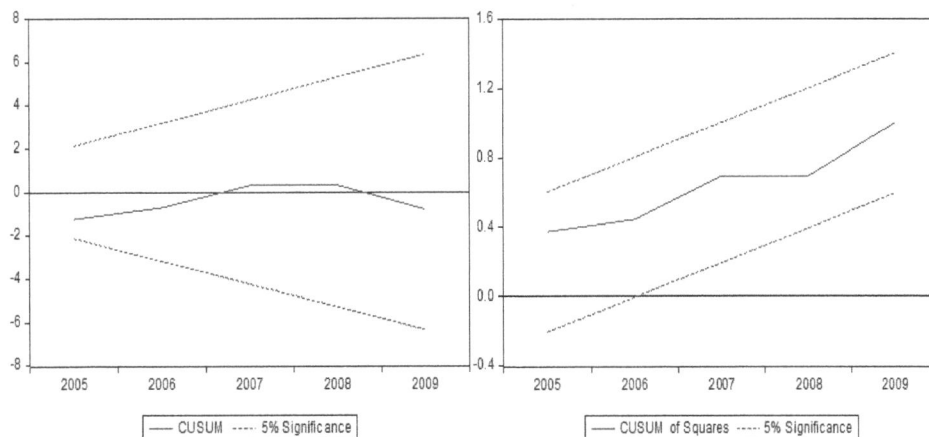

Figure 3. Plot of the CUSUM and CUSUMSQ statistics for a variable LCP

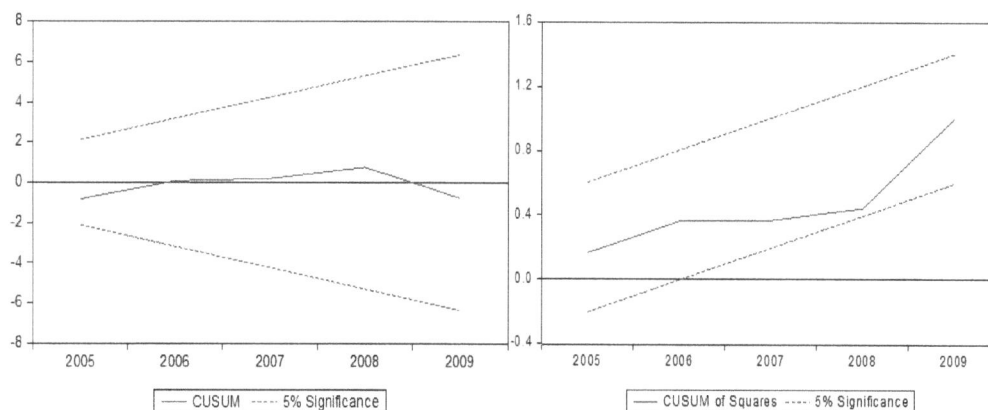

Figure 4. Plot of the CUSUM and CUSUMSQ statistics for a variable LCP

4. Discussion

The results of the study indicate that among the main sources of air pollution, coal production has the biggest impact on the growing number of respiratory diseases in Kazakhstan. Furthermore, it found strong short-run and long-run Granger causality running from coal production to respiratory diseases. The results of the study could be used to better define and incorporate relevant environmental policies into Kazakhstan's development strategies.

All policies and initiatives in Kazakhstan, aimed at reducing environmental burden and promoting renewable energy technologies, focus on recent commitments to reduce greenhouse gases (GHG) emissions and the upcoming EXPO 2017. This combination creates an incompatibility between expansion of the coal production and attempts to meet the emissions reduction target. Both reduction of the coal dependency and promotion of green energy are not only necessary to fulfill the GHG commitment, but also vital to health and well-being of the nation. Currently, coal is the cheapest fuel in Kazakhstan, and therefore the most attractive for both industrial and residential users. However, the health costs associated with the use of coal are often unseen, and if these costs are accounted for, coal becomes an expensive fuel.

This is a first attempt to define the relationship between the air pollution caused by coal production and respiratory health in Kazakhstan at a national level. This study should not be treated as epidemiology research on health effects of air pollution. Unlike existing epidemiology studies on the topic, this study does not intend to estimate effects of various air pollutants (particulate matters, ozone or any other pollutant) on health outcomes (health admissions or deaths). The study investigates long-term equilibrium and short-term dynamics of coal production and respiratory diseases in Kazakhstan.

The results of the research prove that the coal industry is not only the main source of global warming pollutants, but also contributes to the environmental burden of respiratory diseases in Kazakhstan. Hence, it verifies the hypothesis of coal industry's significant contribution to overall health risk attributed to air pollution proposed by Kenessariyev et al. (2013). Similar to the study conducted by Kenessariyev et al. (2013), this study also contains a relatively large uncertainty. However, the history of environmental problems in Kazakhstan and the results of the studies call to act based on the precautionary principle – in the absence of scientific certainty of damage of the action, the burden of the proof should be on the industry that stands to profit (Martuzzi & Tickner, 2004).

Future studies should include analysis on a regional level, given the size of the country and diversity of fuel mix. National level studies provide a broader picture of the issue, however, much of the regional variations remain hidden. Hence, the future studies should include panel data analysis of relationships between fossil fuels combustion and respiratory diseases for all 14 regions of Kazakhstan.

Acknowledgments

The author would like to thank Foundation on Education Development from Yenbekshikazakh district of Almaty region for a support with research.

References

Akhmetov, A., Uchiyama, Y., & Okajima, K. (2012). Life cycle assessment of electricity generation in Kazakhstan. *Energy and Fuel Resources of Kazakhstan*, March/June Issue, 112-114.

Bowen, W. M., Salling, M. J., Haynes, K. E., & Cyran, E. J. (1995). Toward environmental justice: Spatial equity in Ohio and Cleveland. *Annals of the Association of American Geographers, 85*(4), 641-663. http://dx.doi.org/10.1111/j.1467-8306.1995.tb01818.x

BP. (2011). *BP statistical review of world energy 1951-2011*. London, UK: BP.

Brown, R., Durbin, J., & Evans, J. (1975). Techniques for testing the constancy of regression relationships over time. *Journal of the Royal Statistical Society, Series B, 37*(2), 149-192.

Bukhman, M. (2003). *Efficient combustion of coal in Kazakhstan with low emissions of harmful substances into the atmosphere* [Technical paper, in Russian]. Retrieved from http://www.unece.org/fileadmin/DAM/ie/capact/ppp/pdfs/buhman_kzkhstn.pdf

Castelden, W. M., Shearman D., Crisp G., & Finch P. (2011). The mining and burning of coal: Effects on health and the environment. *The Medical Journal of Australia, 195*(6), 333-335. http://dx.doi.org/10.5694/mja11.10169

Concept of Coal Industry Development of Kazakhstan until 2020 [Government resolution, in Russian] (2008, June 28). No. 2008/644.

Dahl, C., & Kuralbayeva K. (2001). Energy and the environment in Kazakhstan. *Energy Policy, 29*(6), 429-440. http://dx.doi.org/10.1016/S0301-4215(00)00137-3

Dasgupta, S., Huq, M., Khaliquzzman, M., Pandey, K., & Wheeler D. (2006). Indoor air quality for poor families: new evidence from Bangladesh. *Indoor Air, 16*(6), 426-444. http://dx.doi.org/10.1111/j.1600-0668.2006.00436.x

Engle, R. F., & Granger, C. W. J. (1987). Co-integration and error correction: representation, estimation and testing. *Econometrica, 55*(2), 251-276.

Epstein, R., Paul, J., Buonocore, J., Eckerle, K., Hendryx, M., Stout III, B. M., ... Glustrom, L. (2011). Full cost accounting for the life cycle of coal. *Annals of the New York Academy of Sciences, 1219*, 73-98. http://dx.doi.org/10.1111/j.1749-6632.2010.05890.x

Finkelman, R., Orem, W., Castranova, V., Tatu, C. A., Belkin, H. E., Zheng, B., ... Bates, A. L. (2002). Health impacts of coal and coal use: possible solutions. *International Journal of Coal Geology, 50*(1-4), 425-443. http://dx.doi.org/10.1016/S0166-5162(02)00125-8

Gelobter, M. (1992). Toward a model of "environmental discrimination". In B. Bryant, & P. Mohai (Eds), *Race and the incidence of environmental hazards: a true time of discourse* (pp. 64-81). Boulder, CO: Westview Press.

Ghose, M. K., & Majee, S. R. (2000). Sources of air pollution due to coal mining and their impacts in Jharia coalfield. *Environment International, 26*(1-2), 81-85.

Gianessi, L. P., Peskin, H. M., & Wolff, E. (1979). The distributional effects of uniform air pollution in the United States. *Quarterly Journal of Economics, 93*(2), 281-301.

Hendryx, M. (2009). Mortality from heart, respiratory, and kidney diseases in coal mining areas of Appalachia. *International Archives of Occupational Environmental Health, 82*, 243-249. http://dx.doi.org/10.1007/s00420-008-0328-y

Hsiao, T. P., Hsiao, C. Y., & Yeou, H. Y. (2011). Modeling the CO2 emissions, energy use, and economic growth in Russia. *Energy, 36*(8), 5094-5100. http://dx.doi.org/10.1016/j.energy.2011.06.004

International Energy Agency (IEA). (2010). *Energy balances of non-OECD countries (2010 Edition)* (p. 177). Paris, France: OECD/IEA.

Johansen, S. (1991). Estimation and hypothesis testing of cointegration vectors in Gaussian vector autoregressive models. *Econometrica, 59*(6), 1551-1580.

Kenessariyev, U., Golub, A., Brody, M., Dosmukhametov, A., Amrin, M., Erzhanova, A., & Kenessary, D. (2013). Human health cost of air pollution in Kazakhstan. *Journal of Environmental Protection, 4*(8), 869-876. http://dx.doi.org/10.4236/jep.2013.48101

Koop, G., McKitrick, R., & Tole, L. (2010). Air pollution, economic activity and respiratory illness: evidence from Canadian cities, 1974-1994. *Environmental Modelling & Software, 25*(7), 873-885. http://dx.doi.org/10.1016/j.envsoft.2010.01.010

Kwiatkowski, D., Phillips, P. C. B., Schmidt, P., & Shin, Y. (1992). Testing null hypothesis of stationarity against the alternative unit root. *Journal of Econometrics, 54*, 159-178.

Lockwood, A. H. (2012). *The silent epidemic: coal and the hidden threat to health.* Cambridge, MA: The MIT Press.

Martuzzi, M., & Tickner J. A. (2004). Introduction – the precautionary principle: protecting public health, the environment and the future of our children. In M. Martuzzi, & J. A. Tickner (Eds.), *The precautionary principle: protecting public health, the environment and the future of our children* (pp. 7-14). Copenhagen, Denmark: World Health Organization Europe.

MedInform Ltd. (2014). Medstat [Medical statistics database, in Russian]. Retrieved from http://medinform.kz/prog.jsp?prog=Medstat

Mennis, J. L. (2005). The distribution and enforcement of air polluting facilities in New Jersey. *The Professional Geographer, 57*(3), 411-422. http://dx.doi.org/10.1111/j.0033-0124.2005.00487.x

Morrice, E., & Colagiuri, R. (2013). Coal mining, social injustice and health: a universal conflict of power and priorities. *Health & Place., 19*, 74-79. http://dx.doi.org/10.1016/j.healthplace.2012.10.006

Narayan, P. K., & Smith, R. (2005). Electricity consumption, employment and real income in Australia evidence from multivariate Granger causality test. *Energy Policy, 33*(9), 1109-1116. http://dx.doi.org/10.1016/j.enpol.2003.11.010

Penney, S., Bell, J., & Balbus, J. (2009). *Estimating the health impacts of coal-fired power plants receiving international financing* [Technical report]. Washington, DC: Environmental Defense Fund. Retrieved from http://www.edf.org/sites/default/files/9553_coal-plants-health-impacts.pdf

Ren, C., & Tong, S. (2008). Health effects of ambient air pollution – recent research development and contemporary methodological challenges. *Environmental Health, 7*(1), 56-66. http://dx.doi.org/10.1186/1476-069X-7-56

Riekert, J. W., & Koch, S. F. (2012). Projecting the external cost of a coal-fired power plant: the case of Kusile. *Journal of Energy in Southern Africa, 23*(4), 52-66. Retrieved from http://www.erc.uct.ac.za/jesa/volume23/23-4jesa-riekert-koch.pdf

Said, S. E., & Dickey, D. A. (1984). Testing for unit roots in autoregressive-moving average models of unknown order. *Biometrika, 71*(3), 599-607.

Sheppard, E., McMaster R. B., Leitner H., Elwood, S., & Gonguo, T. (1999). Examining environmental equity in Hennepin county in Minneapolis. *CURA Reporter, 29*(3), 1-8.

Smith, K. R., & Mehta, S. (2003). The burden of disease from indoor air pollution in developing countries: Comparison of estimates. *International Journal of Hygiene and Environmental Health, 206*(4-5), 279-289. http://dx.doi.org/10.1078/1438-4639-00224

Sobral, H. R. (1989). Air pollution and respiratory diseases in children in Sao Paolo, Brazil. *Social Science & Medicine, 29*(8), 959-964. http://dx.doi.org/10.1016/0277-9536(89)90051-8

Terekhin, S. P., & Pichkhadze, G. M. (1991). The prevalence of primary hypo-vitaminosis among steelmakers and miners of Karaganda region [in Russian]. *Health in Kazakhstan, 3*, 25-26.

The Agency of Statistics of the Republic of Kazakhstan. (2011). *Kazakhstan in the years of independence 1991-2010: statistical compendium* [in Russian]. Astana: The Agency of Statistics of the Republic of Kazakhstan.

World Health Organization (WHO). (2009, February). *Death and DALY estimates for 2004 by cause for WHO member states. Persons, all ages* [Excel spreadsheet]. Retrieved from http://www.who.int/healthinfo/global_burden_disease/estimates_country/en/

Zubov, A. (2007, September 28). Kazakhstan becomes a zone of high ecological risk [in Russian]. *Delovoi Kazakhstan, 37*, 4.

8

On the Integrated Usage of Atomic and Nuclear Approaches for Detecting Potentially Hazardous Elements*

Ashraf S. Elkady[1,2], Walaa M. Abdel-Aziz[2] & Ibrahim I. Bashter[3]

[1] Department of Physics, Faculty of Science for Girls, King Abdulaziz University, Jeddah, KSA

[2] Department Reactor Physics, NRC, Egyptian Atomic Energy Authority (EAEA), Cairo, Egypt

[3] Department of Physics, Faculty of Science, Zagazig University, Zagazig, Egypt

Correspondence: Ashraf S. Elkady, Physics Department, Faculty of Sciences, King Abdulaziz University, KSA. E-mail: elkady8@gmail.com

* Dedicated as a memorial to Professor Abdel-Monem Hassan, EAEA, Egypt, who passed away on the 2nd of January 2011

Abstract

In this work, we report on the usage of different atomic and nuclear approaches for detecting hazardous elements in some commercially available eye cosmetics. Recent studies showed that some eye cosmetics (e.g. eye-liners like kohl) might have hazardous and toxic elements in its elemental composition, which would harmfully impact on the environment and health of its users. In order to obtain accurate information on the elemental content of some natural and synthetic eye-liners that are commercially available in the Egyptian markets and pharmacies, we have applied Energy Dispersive X-ray (EDX), Atomic Absorption Mass Spectroscopy (AA-MS), Elemental Analyzer (EA), and Neutron Activation Analysis (NAA). Heavy and toxic elements, as well as short and long lived radionuclides concentration values were identified in the studied samples. The results indicate that among the three studied samples, the highest lead containing sample is the natural unprocessed one of African source; while the most abundant element in the synthetic samples made in France and USA is Carbon. The present study raises a concern about the medical and environmental implications of using eye-liners, and emphasizes the vital role played by atomic and nuclear approaches in detecting hazardous elements in such commercially available products.

Keywords: atomic and nuclear approaches, hazardous elements, eye cosmetics, energy dispersive x-ray spectroscopy, neutron activation analysis

1. Introduction

Traditional medicine plays an important role and has a direct impact on the general state of population health (Lekounch et al., 2001). However, many medical remedies and mixtures used in this medicine can present a health risk due to the presence of toxic products such as trace elements of lead, cadmium, mercury…etc (Al-Saleh & Coat, 1995). The usage of traditional cosmetics and remedies such as kohl, henna, teething powder… etc, are very common, especially among women, children and babies (Khassouni, 1993).

Kohl is traditional eyeliner that has been widely used as an eye cosmetic in the Middle East, Far East, and Northern Africa (Al-Hawi, 1980). It is both used for beautification and as a traditional ethnic remedy to relieve eyestrain, pain, or soreness. In addition, kohl is known to prevent sun glare, thus it was used by Bedouins in the Arab peninsula. Previous studies (Al-Hazzaa, 1995; Hardy et al., 2004) have shown that kohl contains toxic heavy metals, such as lead, and case studies have revealed that blood lead levels were significantly higher in individuals who used kohl compared to ones who did not (Al-Ashban et al., 2004).

The extensive use of green, white and black make-up has been known since the earliest periods of Egyptian history (Walter et al., 1999). Recently, some authors have suggested the potential adverse effects of some eye make-ups on health, such as lead poisoning that makes them a health risk and of worth investigation in different countries (Carol & Joseph, 1991; Hardy et al., 2004; Karim et al., 2005). Besides, a diverse array of nanomaterials including nanocosmetics, has become widespread in use due to the fast development of nanotechnology applications. However, the safety of such materials has not been well assessed, because they have been considered as safe as common larger sized materials, which are known not to be absorbed by the body (Yoshida et al., 2012). Indeed, this is not true due to the fact that when the cosmetic particles size is reduced down to nanoscale, they become more able to penetrate cellular and brain barrier membranes. Thus, the wide use of very fine particles of kohl on the eyes of children and adults motivated us to investigate whether kohl is a possible source of exposure to lead and other toxic elements in Egypt.

In this context, there are different atomic and nuclear techniques that can be used for elemental analysis and for detecting toxic and hazardous heavy elements. However, each technique has its detection limit, accuracy and sensitivity to certain elements. Neutron activation analysis (NAA) is considered one of the most important analytical nuclear approaches, which yield a precise result for trace and ultra-trace elemental concentrations in complex samples. Along the past several decades, it was successfully applied for determination of a great variety of elements in many disciplines including environmental (Landsberger, 1992), biological (Erdtmann & Petri, 1986), geological as well as material sciences (Martinez, 1997). It is considered as a method for qualitative and quantitative determination of elements based on the measurement of characteristic radiation from radionuclides formed directly, or indirectly by neutron irradiation of samples (Hassan et al., 1994). The most suitable source for neutrons to be used in this technique is usually a nuclear research reactor. The high resolution gamma-ray detection systems are used for analysis of the complex gamma-ray spectra obtained by neutron capture.

In the present study, we applied different atomic and nuclear approaches, namely Energy Dispersive X-ray (EDX), Atomic Absorption Mass Spectroscopy (AA-MS), Elemental Analyzer (EA) and Neutron Activation Analysis (NAA) in order to obtain more information on the composition and elemental content of some natural and synthetic eye-liner samples taken from the Egyptian markets and pharmacies. The emphasis will be on the medicinal implications of using eye-liners with high content of lead and toxic elements, and on the vital role that atomic and nuclear methods play in detecting such hazardous elements in eye-cosmetics.

2. Materials and Methods

2.1 Samples Preparation

Samples of kohl used in this work were purchased from local Egyptian markets and pharmacies in Cairo. Three commonly used representative samples were chosen for analysis. The African natural untreated eye-liner was in big crystallized stone form, while the American and French synthetic kohl were in pencil forms. The samples were crushed, finely grounded into powder form before investigations.

2.2 Experimental Approaches

Different atomic and nuclear approaches were applied in order to obtain a complete picture of the elemental content in the studied samples, and avoid the limitation and accuracy of a single approach.

Atomic Absorption Spectrometer, Energy Dispersive X-ray spectrometer, and Analytical analyzer located in the Central Laboratory of the Nuclear Research Center (NRC), Egypt were used to identify the chemical elements present in the samples under investigation. Briefly, a flame atomic absorption spectrometer was used in order to analyze the samples for their elemental constituents. The samples were atomized and optically irradiated; the radiation then passes through a monochromator in order to separate the element-specific radiation from any other radiation emitted by the radiation source, which is finally measured by a detector.

In EDX facility, primary x-rays are illuminated from the x-ray tube to the specimen, fluorescence x-rays having wavelengths (energies) peculiar to the constituent elements of the specimen are generated from the elements. Qualitative analysis can be made by investigating the wavelengths of the fluorescence x-rays and quantitative analysis by investigating the x-ray dose. The energies are investigated by using the energy separation characteristic of x-ray detector (Willard et al., 1988).

The elemental analyzer detector used in this study is a thermal conductivity detector (TCD) sensitive to any substance with thermal conductivity other than that of the carrier gas. The detector consists of a stainless steel block provided with two pairs of filaments (of tungsten rhenium) having the same electrical resistance. The detector is housed in a thermally insulated metal block (detector oven) and maintained at constant temperature. The two pairs of filaments are electrically connected according to a Wheatstone bridge circuit powered at constant voltage. The first pair of filaments are fed with pure carrier gas (reference channel), where the second pair are fed with the gas flowing from the reactor (analytical channel). When the bridge is powered, the filaments are heated at a temperature (resistance) that is a function of the thermal conductivity of the gas feeding the filaments. The reference channel is exposed only to pure carrier gas, whereas the analytical channel is exposed to the reactor effluents (carrier gas + sample).

When pure carrier gas flows through the reference and the analytical channels, a constant temperature gradient is established between the elements and the detector walls and the Wheatstone bridge is balanced where there is no output signal. As a component is eluted, a change in heat transfer occurs, with consequent variation of the filaments temperature. Since electric resistance is a function of temperature, the bridge unbalances and the detector generates a signal proportional to the difference in thermal conductivity between the eluted component and the carrier gas. The output signal is then sent to the data acquisition board.

Neutron Activation Analysis (NAA) technique, available at the ET-RR-2 reactor was applied to the natural eye-liner sample that was prepared in dried powder form. The other two samples were too brittle to be irradiated by high flux of neutrons. Briefly, neutrons as well as charged particles can react with isotopes of various elements and produce radioactive nuclides. The characteristic radiation emitted by these nuclides can be used for qualitative and quantitative determination of various elements. In this way, elements in part per million or percentage can be analyzed. Usually, neutrons are used as projectiles and γ-rays are emitted. Also the high resolution γ-ray detection system together with the advanced computer programs can help in complete analysis of the data obtained with high accuracy. Neutron fluxes in the order of 1.3×10^{11} and 2.7×10^{11} n.cm^{-2}.s^{-1} were applied for long and short irradiation times respectively during sample irradiation.

For qualitative and quantitative analysis of the minor elements in sample 3, the well-resolved and pronounced γ-ray lines in the obtained γ spectra have been selected, and the well-known analytical equation was used (Nada et al., 2001):

$$m = \frac{\lambda\, C\, M}{\left[\left(\varepsilon I_r \sigma_{th} f N_0 \phi_m\right) e^{-\lambda t_W} \left(1 - e^{-\lambda t_{irr}}\right)\left(1 - e^{-\lambda t_C}\right)\right]}$$

where m is the mass of the element; ϕ_m is the thermal neutron flux; λ is the decay constant; C is the activity (net peak area of gamma-ray line); M is the atomic mass; ε is the efficiency of the detection system at the selected full energy peak, I_γ is the absolute intensity of the gamma-ray line; σ_{th} is the thermal (n, γ) cross-section; f is the isotopic abundance fraction; t_w is the cooling time; t_{irr} is the irradiation time, t_c is the counting time; and N_0 is the Avogadro's number. The main factors used in the calculation were the isotopic abundance of the selected isotopes for each element, half-life, cross-section of the (n, γ) reaction and the intensity of the selected γ-ray line (Lederer & Shirley, 1997).

3. Results and Discussions

3.1 EDX, AA-MS and EA

The results obtained from EDX measurements are given in Table 1 and Figure 1. It is noticeable that the most predominant elements are C, Cu, Al and Si in sample 1; C, O, Si, Fe and Zn in sample 2, while the concentrations of Pb and S in sample 3 are considerably high.

Table 1. Elements concentration (%) in the three samples of eye-cosmetics as detected by EDX

Element	Sample 1 (France)	Sample 2 (USA)	Sample 3 (Africa)
C	94.29± 4.7145	52.14± 2.607	ND
O	ND	16.83± 0.8415	ND
Al	1.04±0.052	1.16± 0.058	0.84±0.042
Si	1.14±0.057	2.74± 0.137	ND
Ca	ND	0.71±0.0355	ND
Mn	ND	0.24±0.012	ND
S	0.05±0.0025	ND	12.34±0.617
K	0.48±0.024	ND	ND
Ti	0.48±0.024	ND	ND
Fe	0.39±0.0195	24.39±1.2195	ND
Cu	1.27±0.0635	0.19±0.0095	ND
Zn	0.87±0.0435	1.59±0.0795	ND
Pb	ND	ND	86.82±4.341

* ND: Not detected.

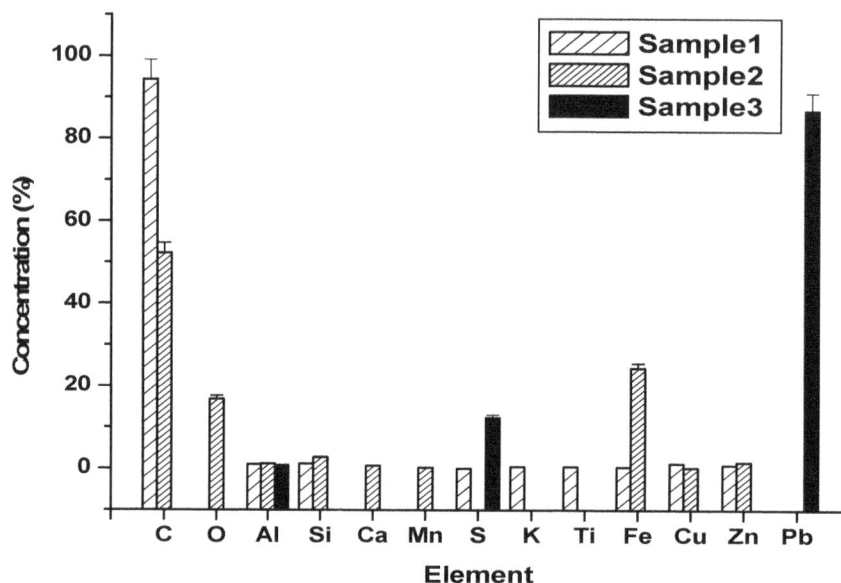

Figure 1. Elements concentration (%) in the three samples of eye-cosmetics, as detected by EDX

The AA-MS results are represented in Figure 2. One can note that the cupper and iron concentrations are comparable among the three studied samples (see Figure 2). However, the heavy element Zn is found in considerable concentration in sample 3 and of lower concentrations in sample 1 and sample 2.

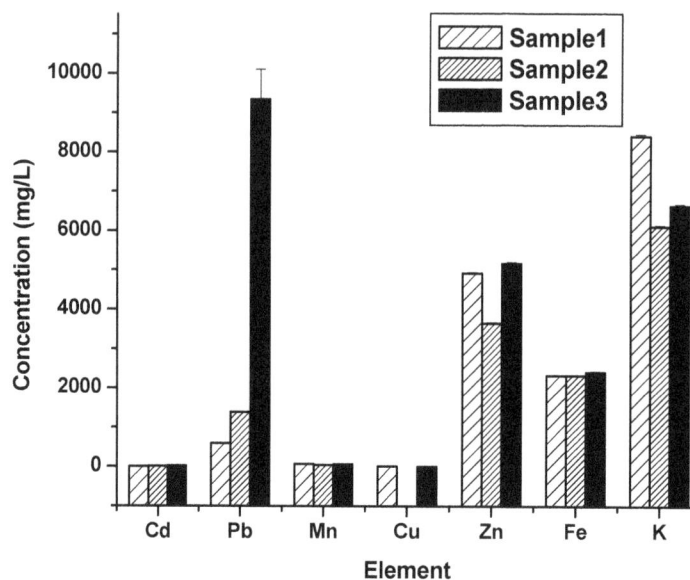

Figure 2. Elements Concentration (mg/L) in the three samples of eye-cosmetics, as measured by AA-MS. The concentration of Pb is divided by 10^2

Therefore, out of the three tested set of samples, one contained in excess of 86% lead (from Africa), the other two samples from France and USA contained about 5% and 13% lead respectively. While the concentration of Al was comparable in the three samples; the Cd concentration was highest in sample 3 (from Africa) and lowest in sample 1 (from France), as indicated from EDX and AA-MS measurements. There were no significant concentration differences among other elements (e.g. Mn, Si and K) in the three samples according to AA-MS measurements, which are also consistent with EDX measurements.

The elemental analyzer measurements showed that the prevailing element in sample 1 is carbon (88.34%), while it is minimal in sample 3 (0.25%). The latter low ratio was not detected by EDX, but was detected by the elemental analyzer technique (see Table 2). There is a noticeable agreement in the existing carbon and sulfur concentrations measured using EDX and elemental analyzer facilities.

Table 2. Concentration of (C, O, S) in the studied samples (%), obtained using EA

Element	Sample 1 (France)	Sample 2 (USA)	Sample 3 (Africa)
C	88.34±4.417	52.07±2.6035	0.2584±0.01292
O	4.896±0.2448	8.733±0.43665	2.2522±0.11261
S	ND	ND	11.23±0.5615

* ND: Not detected.

From the above described results, it is worth noting that among the detected elements in the composition of samples is lead, which constitutes the main hazardous element. The highest concentration of lead is found in the natural, unprocessed sample from African sources (sample 3), while its concentration was significantly reduced in the other two samples, manufactured in France and USA. Thus, the chemical treatment during manufacture would have its impact on lead content of the processed product.

It is well established that lead is harmful to all adults, children and infants. It is very toxic and can be introduced into the body by ingestion, inhalation, and by skin exposure. Lead poisoning is a global problem, considered to be the most important environmental disease in children (Tong et al., 2000). It is particularly harmful to the developing brain and nervous system. Eye cosmetics like kohl, with high lead concentration, are frequently used by women and as a skin treatment product on infants. High blood lead levels in kohl-using infants and in infants of kohl-using mothers have been reported (5.2μg/dl versus 2.8μg/dl) and considered abnormal (Nir et al., 1992; Klaasen, 1996; Ahmed et al., 2005). Such abnormal blood lead levels are considered unsafe and have been shown to be associated with intelligence quotient deficits, behavioral disorders, slowed growth, and impaired hearings (Karri et al., 2008; Needleman et al., 2000; Schwartz & Otto, 1991).

3.2 Neutron Activation Analysis

The concentration of minor elements in sample 3 as estimated from γ-ray spectra, obtained using the short and long irradiation facilities, are given in Tables 3, 4 respectively, and represented in Figure 3. It is noticeable that among the 8 detected elements by NAA, there are 7 heavy elements, namely Ag, Sb, Mn, Cd, Zn, V and Th. Of worth noting, the Al and Cd concentration values obtained from NAA are consistent with their corresponding values obtained using EDX and AA-MS respectively.

Table 3. The heavy elements (except for Al) concentrations in sample 3 as obtained from NAA, using the short irradiation facility

Element	Concentration (ppm)
Al	7930
V	1.700
Th	0.120
Mn	3145

Table 4. The heavy elements concentration in sample 3 as obtained from NAA, using the long irradiation facility

Element	Concentration (ppm)
Ag	11831.7
Sb	7145.30
Cd	23.83
Zn	6.8200

In addition to lead, aluminum might be also toxic at both environmental and therapeutic levels. Aluminum exposure, apart from causing cholinotoxicity, can induce changes in other neurotransmitter levels since neurotransmitter levels are closely interrelated (Bernardi, 2002). Besides, other adverse developmental effects of aluminum on children and infants were previously reported (Al-Saleh & Shinwari, 1996).

Cadmium is also considered one of the extremely toxic heavy metals, which can accumulate with time in the body. Due to its low permissible exposure limit, overexposures may occur even in situations where trace quantities of cadmium are found. There are adverse implications due to exposure to environmental Cd, and it has been linked with e.g. an increased risk of dental caries (Amr & Helal, 2010). Importantly, compounds containing lead and cadmium are also considered potential human carcinogens. However, the mechanism of metal-induced carcinogenesis is still unknown, but one possible pathway may involve the interaction of metals with DNA, either directly or indirectly (Valverde et al., 2001).

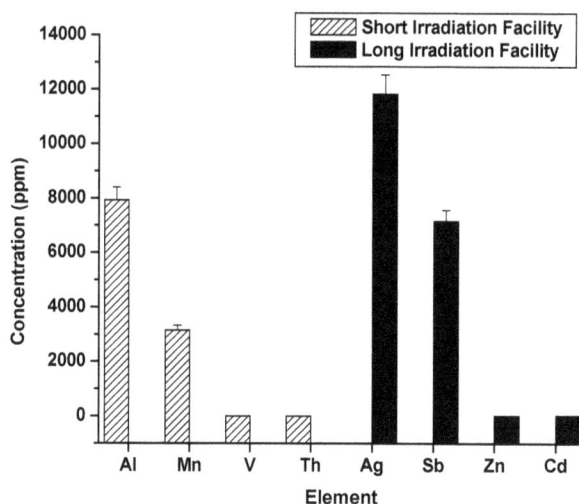

Figure 3. Heavy elements and aluminum concentrations (ppm) in sample 3 as determined by NAA

Besides, Antimony and many of its compounds are toxic, and the effects of antimony poisoning are similar to arsenic poisoning. It has also been found to induce DNA strand lesions but not DNA-protein crosslinks (Mourón et al., 2006). Acute (short-term) exposure to antimony by inhalation in humans results in effects on the skin and eyes. Respiratory effects, such as inflammation of the lungs, chronic bronchitis, and chronic emphysema, are the primary effects noted from chronic (long-term) exposure to antimony in humans via inhalation. Human studies are inconclusive regarding antimony exposure and cancer, while animal studies have reported lung tumors in rats exposed to antimony trioxide via inhalation (Sundar & Chakravarty, 2010).

On the other hand, silver is intimately associated with environmental contamination of other toxic heavy metals such as mercury and lead. It does not play a known natural biological role in humans, and possible health effects of silver are a disputed subject. Silver itself is not toxic to humans, but most silver salts are. In large doses, silver and compounds containing it can be absorbed into the circulatory system and become deposited in various body tissues, leading to Argyria, which results in a blue-grayish pigmentation of the skin, eyes, and mucous membranes (Hammond, 2000). Interestingly, comparative studies with both silver ions (such as silver acetate) and polyvinylpyrrolidone (PVP)-stabilized silver nanoparticles (70 nm) showed that the effective toxic concentration of silver towards bacteria and human cells is almost the same (Greulich et al., 2012).

Furthermore, vanadium (V) might hold a toxic potential to human health and the environment, especially if the kohl particles are ultra-fine down to nanosize. Due to the high surface-to-volume ratio, small amounts can lead to strong oxidative damage within biological systems, impairing cellular functions as a consequence of their surface reactivity. Recent studies showed that nanoparticulate vanadium oxide has potential vanadium toxicity in human lung cells (Wörle-Knirsch et al., 2007).

Finally, it is worth mentioning here that there is evidence to indicate that, compounds used in cosmetics like galena, are likely to form nanocrystals within e.g. hair during blackening (Walter et al., 2006). In our viewpoint, such tiny crystals would much influence the toxicity in living tissues due to their expected higher penetration to cellular and brain barrier membranes. The nanotoxicology of such nanocrystals is an important issue that worth further investigations. In this context, the human health effects resulting from exposure to nanocosmetics should be better explored in order to understand the possible potential hazards of these nanomaterials that already exist in the global market (Bowman & Van Calster, 2008).

4. Conclusions

We have applied an integrated approach by using different atomic and nuclear techniques for probing hazardous elements in some eye-cosmetics, commercially available in Egypt. The applied combined approach allowed for qualitative and quantitative identification of almost the entire elemental content in the studied samples, and helped to overcome the limitation and accuracy of using a single technique. A total of three synthetic and natural eye-liner samples of known origin were analyzed using Energy Dispersive X-ray (EDX), Atomic Absorption Mass Spectroscopy (AA-MS), Elemental Analyzer (EA) and Neutron Activation Analysis (NAA). It was found that lead (>86%) represents the main abundant hazardous element in the natural eye-liner sample from African sources. Al and heavy metals like Ag, Sb, Mn, Cd, Zn, V and Th were also identified in the later sample using NAA. For the synthetic two samples from French and American sources, the major hazardous element found to be Carbon in high concentration 94% and 52% respectively. Producing such eye-cosmetics at the nanoscale would increase their harmful effects due to the ability of toxic nanoparticles to cross the cellular and brain barrier membranes. Whence, the present study highlights the medical and environmental implications of using eye cosmetics with high content of lead and heavy toxic elements. Moreover, the ethical and medical concerns raised in this study emphasize the need for global regulatory rules for the cosmetic industry.

Acknowledgements

The authors would like to express their thanks and appreciation to Professor Naguib Ashoub, and Dr. Abd El-Ghany El Abd, EAEA, Egypt for their useful comments and discussions.

References

Ahamed, M., Verma, S., Kumar, A., & Siddiqui, M. K. J. (2005). Environmental exposure to lead and its correlation with biochemical indices in children. *Science of the total environment, 346*(1), 48-55. http://dx.doi.org/10.1016/j.scitotenv.2004.12.019.

Al-Ashban, R. M., Aslam, M., & Shah, A. H. (2004). Kohl (surma): a toxic traditional eye cosmetic study in Saudi Arabia. *Public Health, 118*(4), 292-298. http://dx.doi.org/10.1016/j.puhe.2003.05.001

Al-Hawi, S. A., Wafai, M. Z., Kalahari, M. R., & AL-Ugum, W. A. (1980). *Light of the Eyes and the Collector of Arts* (Vol. 1286, p. 142). King Faisal Center for Research and Islamic Studies, Saudi Arabia.

Al-Hazzaa, S. A., & Krahn, P. M. (1995). Kohl: a hazardous eyeliner. *Inter Ophthal., 19*, 83-88. http://dx.doi.org/10.1007/BF00133177

Al-Saleh, I., & Coat, L. (1995). Lead exposure in Saudi Arabia from the use of traditional cosmetics and medical remedies. *Environ. Geochem. Health, 17*, 29-31.

Al-Saleh, I., & Shinwari, N. (1996). Aluminum in Saudi children. *Biometals, 9*, 385-392. http://dx.doi.org/10.1007/BF00140608

Amr, M. A., & Helal, A. I. (2010). Analysis of trace elements in teeth by ICP-MS implication for caries. *Journal of physical science, 21*(2), 1-12.

Bernardi, R. A. (2002). *Protect children from the dangers of lead.* HUD announces $10 million to help communities. HUD no. 02-022. Washington: US Department of Housing and Urban Development.

Bowman, D. M., & Van Calster, G. (2008). Flawless or Fallible? A Review of the Applicability of the European Union's Cosmetics Directive in Relation to Nano-Cosmetics. *Studies in Ethics, Law, and Technology, 2*(3).

Erdtmann, G., & Petri, H. (1986). Nuclear activation analysis: fundamentals and techniques. In P. J. Elving (Ed.), *Treatise of Analytical Chemistry* (2nd Ed., p. 414). New York: Wiley.

Greulich, C., Braun, D., Peetsch, A., Diendorf, J., Siebers, B., Epple, M., & Köller, M. (2012). The toxic effect of silver ions and silver nanoparticles towards bacteria and human cells occurs in the same concentration range. *RSC Advances, 2*(17), 6981-6987. http://dx.doi.org/10.1039/c2ra20684f

Hammond, C. R. (2000). *The Elements, in Handbook of Chemistry and Physics 81st edition.* CRC press. ISBN 0-8493-0481-4.

Hardy, A. D., Walton, R. I., & Vaishnave, R. (2004). Composition of eye cosmetics (kohls) used in Cairo. *J. Enviiron. Health Res., 14*(1), 83-91. http://dx.doi.org/10.1080/09603120310001633859

Hassan, A. M., El-Enany, N., El-Tanahy, Z., & Abdel-Momen, M. A. (1994). Multielement measurements of various industrial samples by neutron capture prompt gamma-ray activation analysis. *Nucl. Geophys., 8*(1), 91-98.

Karim, N. J., & Hartmut, G. H. (2005). Characterization of a hazardous eyeliner (kohl) by confocal Raman microscopy. *Jounal of Hazardous Materials, B124*, 236-240.

Karri, S., Saper, R., & Kales, S. (2008). Lead Encephalopathy Due to Traditional Medicines. *Curr Drug Saf., 3*(1), 54-59. http://dx.doi.org/10.2174/157488608783333907

Khassouni, S. (1993). *Utilisation du khol: risqué d'intoxication* (3rd Ed., p. 188).

Klaasen, C. D. (1996). *Casarett and Doull's Toxicology, the Basic Science of Poisons* (5th ed.). New York: McGraw-Hill Book Company.

Landsberger, S. J. (1992). Analytical methodologies for instrumental neutron activation analysis of airborne particulate matter. *J. Trace Microprobe Tech., 10*, 1.

Lederer, C. M., & Shirley, V. S. (1997). *Table of isotopes* (8th ed.). New York: Wiley.

Lekouch, N., Sedki, A., Nejmeddine, A., & Gamon, S. (2001). Lead and traditional Moroccan pharmacopoein. *The Science of the Total Environment, 280*, 39-34. http://dx.doi.org/10.1016/S0048-9697(01)00801-4

Martinez, T., Lartigue, J., Navarrete, M., Avila, P., Lopez, C., Cabrera, L., & Vilchis, V. (1997). Determination of pollutants in dwellings by neutron activation analysis and X-ray fluorescence. *J. Radioanal. Nucl. Chem., 216*, 37-39. http://dx.doi.org/10.1007/BF02034492

Mourón, S., Grillo, C., Dulout, F., & Golijow, C. (2006). Induction of DNA strand breaks, DNA-protein crosslinks and sister chromatid exchanges by arsenite in a human lung cell line. *Toxicology in Vitro, 20*(3), 279-285. http://dx.doi.org/10.1016/j.tiv.2005.07.005.

Nada, A., El-Bahi, S. M., Abdel-Ghany, H. A., & Hassan, A. M. (2001). Elemental investigation of some Egyptian vehicle motor alloys. *Applied Radiation and Isotopes, 55*, 575-580. http://dx.doi.org/10.1016/S0969-8043(01)00099-9

Needleman, H. L., Schell, A., Bellinger, D., & Alfred, E. N. (1990). The long-term effects of exposure to low doses of lead in childhood. An eleven-year follow-up report. *N. Engl. J. Med., 332*, 83. http://dx.doi.org/10.1056/NEJM199001113220203

Nir, A., Tamir, A., Zelmuk, A., & Iancu, T. C. (1992). Is eye cosmetic a source of lead poisoning. *Isr. J. Med. Sci., 28*(7), 417.

Parry, C., & Eaton, J. (1991). Kohl: a lead-hazardous eye makeup from the Third World to the First World. *Environmental Health Perspectives, 94*, 121. http://dx.doi.org/10.2307/3431304.

Schwartz, J., & Otto, D. (1991). Lead and minor hearing impairment. *Archives of Environmental Health: An International Journal, 46*(5), 300-305. http://dx.doi.org/10.1080/00039896.1991.9934391

Sundar, S., & Chakravarty, J. (2010). Antimony toxicity. *International journal of environmental research and public health, 7*(12), 4267-4277. http://dx.doi.org/10.3390/ijerph7124267

Tong, S., von Schirnding, Y. E., & Prapamontol, T. (2000). Environmental lead exposure: a public problem of global dimension. *Bulletin of the World Health Organization, 78*, 1068-1077. http://dx.doi.org/10.1590/S0042-96862000000900003

Valverde, M., Trejo, C., & Rojas, E. (2001). Is the capacity of lead acetate and cadmium chloride to induce genotoxic damage due to direct DNA–metal interaction? *Mutagenesis, 16*(3), 265-270. http://dx.doi.org/10.1093/mutage/16.3.265

Walter, P., Martinetto, P., Tsoucaris, G., Brniaux, R., Lefebvre, M. A., Richard, G., ... Dooryhee, E. (1999). Making make-up in Ancient Egypt. *Nature, 397*(6719), 483-484. http://dx.doi.org/10.1038/17240

Walter, P., Welcomme, E., Hallégot, P., Zaluzec, N. J., Deeb, C., Castaing, J., ... Tsoucaris, G. (2006). Early use of PbS nanotechnology for an ancient hair dyeing formula. *Nano letters, 6*(10), 2215-2219. http://dx.doi.org/10.1021/nl061493u

Willard, H. H., Merritt Jr, L. L., Dean, J. A., & Settle Jr, F. A. (1988). *Instrumental methods of analysis.* USA:

Wadsworth.

Wörle-Knirsch, J. M., Kern, K., Schleh, C., Adelhelm, C., Feldmann, C., & Krug, H. F. (2007). Nanoparticulate vanadium oxide potentiated vanadium toxicity in human lung cells. *Environmental science & technology, 41*(1), 331-336. http://dx.doi.org/10.1021/es061140x

Yoshida, T., Yoshioka, Y., & Tsutsumi, Y. (2011). [The safety assessment of nanomaterials for development of nano-cosmetics]. *Yakugaku zasshi: Journal of the Pharmaceutical Society of Japan, 132*(11), 1231-1236. http://dx.doi.org/10.1248/yakushi.12-00232-4

9

Application of Waste Plastic Pyrolysis Oil in a Direct Injection Diesel Engine: For a Small Scale Non-Grid Electrification

Sunbong Lee[1], Koji Yoshida[2] & Kunio Yoshikawa[1,2]

[1] Department of Environmental Science and Technology, Interdisciplinary Graduate School of Science and Engineering, Tokyo Institute of Technology, Kanagawa, Japan

[2] Department of Mechanical Engineering, College of Science and Technology, Nihon University, Tokyo, Japan

Correspondence: Sunbong Lee, Department of Environmental Science and Technology, Tokyo Institute of Technology, Kanagawa, Japan. E-mail: wte.sblee77@gmail.com

Abstract

Waste plastic can be transformed to oil by the pyrolysis and it may be applicable as a fuel for diesel engines. The pyrolysis oil property varies depending on the raw waste plastic and the pyrolysis condition, which is different from that of diesel and gasoline. Considering the thermal efficiency, the running stability and the reliability, diesel engines are the most promising energy converter to generate electricity by using the pyrolysis oil. In this research, plastics from municipal wastes were converted into oil through the pyrolysis and the catalytic reforming process in a commercial facility. Compared with diesel fuel, the raw pyrolysis oil showed slightly lower kinematic viscosity than the minimum level of diesel fuel and almost the same heating value. Its carbon class differed from diesel, gasoline and kerosene and is mainly composed of naphethenes and olefins which have poor self-ignition quality. The pyrolysis oil was blended with diesel fuel with different mixing ratios. A single cylinder small size direct injection diesel engine was used for the test. The full load performance, the exhaust emissions and the thermal efficiency were investigated from the view point of the compatibility to diesel fuel based on the US EPA regulation mode.

Keywords: waste plastic, pyrolysis, blend, direct injection diesel engine, exhaust emissions

1. Introduction

The population which could not access to electricity was around 1.2 billion in 2010 and is distributed in many low developing countries such as African, South Central Asia and Southeast Asia/Pacific (World Energy Council, 2013). With the increase in the population and the economic growth in those countries (United Nations, 2012), waste generation is growing rapid especially for the organic and the plastic and the uncontrolled waste disposal is becoming more serious issues to manage it (Daniel, H., 2012). The interest on Waste to Energy is growing by the above drivers (Luca, Lo Re L., 2013). Among them, waste plastic can be converted into oil and gas by the pyrolysis and the catalytic reforming process. The pyrolysis oil has almost the same heating value with petroleum fuels because plastic material is originated from crude oil. The chemical and physical properties of the pyrolysis oil are highly depending on the raw waste plastic. Considering the various oil properties, diesel engine is the most appropriate energy converter in the point of stable engine running as well as the high electricity generation efficiency. Actually, it is unclear whether the pyrolysis oil can be used in diesel engines as 100% alternative fuel or it needs blending with diesel fuel.

The oil from the scrap-tire thermal-mechanical pyrolysis was run on a single cylinder diesel engine using automotive diesel fuel and two mixtures of the diesel fuel and the tire pyrolysis oil (TPO) with the volume ratio of 20% TPO and 40% TPO due to the lower cetane index of TPO. The 20% mixture of TPO and the diesel fuel showed no significant differences for power, fuel consumption and exhaust gas emissions (Stefano, F., 2013). The shredded PVC (Polyvinyl chloride) and PET (Polyethyelene terephthalate) bottle plastics was pyrolyzed in the laboratory scale pyrolysis device and the pyrolysis oil was mixed with diesel fuel using a ultrasonic vibrator due to the high kinematic viscosity around 3 times than that of diesel. The blend oil of diesel with 5% volume fraction ot the waste plastic pyrolysis oil was tested in a single cylinder diesel engine at 50% load of the engine maximum power output and resulted in a slight improvement of the fuel consumption (Senthilkumar, T., 2012). Plastic waste was converted into liquid oil by using the pyrolysis process including the catalytic cracking with the pre-treatment of the shredding. A blend of 50% and 70% with the pyrolysis oil by volume in the diesel fuel were evaluated for the thermal efficiency and the exhaust gas emissions by operating a single cylinder diesel engine at its maximum engine output speed with varying the engine load and showed the higher thermal efficiency and the higher gas emission for CO (Carbon monoxide) and THC (Total hydroCarbon) emission (Rajesh, G., 2011). Mukherjee et al (Kaustav, M., 2014) produced the waste plastic oil via the pyrolyzer without the catalytic process and blended with the diesel fuel and the ethanol by the ratio of 20, 40 and 60% and tested in a direct injection twin cylinder diesel engine at the rated engine speed by changing the engine load. They concluded that the 20% blend oil has characteristics as close to diesel regarding to the thermal efficiency, NOx (Nitrogen oxides) and CO emission. Pratoomyod et al (Jane, P., 2013) carried out the operation test in a direct injection 6-cylinder diesel engine by using the blend oil which was diesel blended with the waste plastic oil with the ratio of 25%, 50% and 75% by volume. The engine operation point was ranged from the two thirds of the rated engine speed to the idle speed at a certain interval on the 100% load. The specific fuel consumption, the CO and the THC emissions were higher than diesel operation and the NO emission increased with the increase in the blend ratio. Several research results were reviewed by Patel et al (Nilamkumar, S, P., 2013) and Harshal et al (Pawar, Harshal R., 2013) about the usage of the blended waste plastic pyrolysis oil with diesel fuel in the diesel engine on the engine operation range which was specified by the researchers.

As reviewed above, the literatures have been highly evaluated academically and been thought to be somewhat limited in the view of the real world adaption of the pyrolysis oil into the diesel engine. The production facility of the raw pyrolysis oil was the laboratory scaled facility or undefined specifically. Engine operation points for the experiment were decided by the view of researchers based on their knowledges and experiences. For the evaluation of the experimental result, the exhaust emissions from the combustion of the pyrolysis oil in the diesel engine were qualitatively compared with those of diesel fuel.

This research was carried out for aiming to the real world adaption mainly for electricity generation. For the sake of it, the pyrolysis oil which was produced in a commercial plant was applied to a small direct injection diesel engine. The compatibility of the pyrolysis oil to diesel fuel was investigated with regard to the full load performance, the exhaust emissions and the thermal efficiency according to the US EPA emission test mode.

2. Materials and Methods

2.1 Production of Waste Plastic Pyrolysis Oil

Municipal solid wastes are collected as shown in Figure 1 and transported to a waste management company. Many kinds of plastic wastes which are containing the residues of food, drinking beverage and water are packed in a plastic bag in the stage of disposal from households. There is no segregation process in the company as pre-process ahead of the pyrolysis process due to the cost.

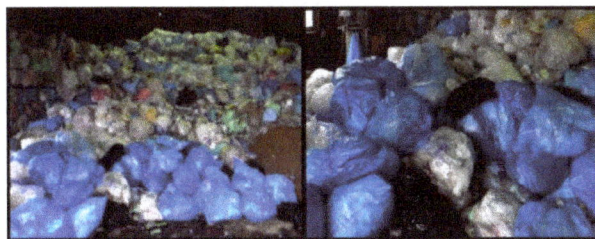

Figure 1. Collection of raw plastic wastes

The conversion of plastics to oil is based on the pyrolysis under the absence of oxygen and the catalytic reforming and the condensation of the resulting gases as illustrated in Figure 2. The semi-batch type reactor which is movable is used. The product gas from the pyrolysis reactor flows into the alumina-based catalytic reformer by the force of vaporizing pressure. The oil burner at the bottom of the jacket surrounding the reactor heats the reactor for the thermal decomposition reaction with the temperature controlled at 400°C and the exhaust gas flows into the outer side jacket of the catalytic reformer, heats the reformer and then exhausted out via the vent. The product gas is further cracked in the catalytic reformer. After passing the reformer, direct scrubbing by spraying the oil is carried out. The condensed oil is then collected in the oil receiver. The off-gas is supplied to the incinerator next to this pyrolysis system to be completely burned. After finishing the pyrolysis reaction in the reactor, an overhead crane lifts up the reactor and installs a new reactor which is already packed with the feedstock and the whole process starts again.

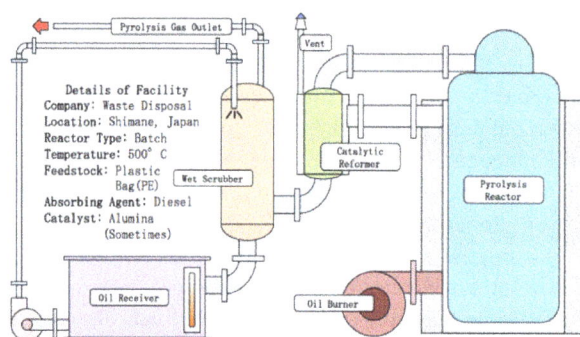

Figure 2. Pyrolysis process of the commercial plant

2.2 Test Fuels

Table 1 shows the hydrocarbon composition analysis and the physical property of the raw pyrolysis oil and diesel fuel, respectively. The raw pyrolysis oil shows almost the same chemical content in carbon and hydrogen except oxygen. It is inferred that the oxygen in the pyrolysis oil came from the contamination of the raw plastic wastes by food and drinking residues in the stage of consumer disposal. The amount of oxygen is low enough and can be neglected as one of the effective factors influencing the engine performances. The density and the kinematic viscosity are lower than those of diesel fuel. The density of the pyrolysis oil may affect the full load performance. The kinematic viscosity undergoes slightly less than the minumue value of the diesel fuel in the United States (ASTM D975). When the raw pyrolysis oil is directly used in diesel engines without any treatments, it would be concerned in terms of the lubricity in the fuel delivery and the injection system of diesel engines (Lacey, P., I., 1992). The heating value is almost the same as diesel due to the similar content of hydrocarbon as stated above.

Table 1. Composition analysis

Composition analysis		
%wt/wt	Diesel [JIS2]	Raw pyrolysis-oil [Waste plastic]
C	85.04	85.87
H	13.55	13.71
N	0	0
S	0	0
O	0	0.421
Physical property		
Density [kg/cm^3]	0.83	0.75
Kinematic viscosity [mm2/s]	2.744	1.19
Water content [%wt/wt]	0	0
Low Heat value [MJ/kg]	45	45.4

Figure 3 illustrates the carbon atom number distribution of the pyrolysis oil obtained from the commercial plant. The raw pyrolysis oil was analyzed by GC-MS analyzer to compare the distribution of hydrocarbon atom to that of diesel. The carbon number distribution of the oil product from the commercial plant was obviously shifted from the hydrocarbon range of diesel. The large fraction of hydrocarbon was found in the range of 6-14, which is much lighter than diesel and is close to the range of gasoline. It reached the maximum fraction at the carbon number of 11. In contrast, the hydrocarbon range of diesel was in the carbon number of 8-25 and the highest fraction is seen at 17. The fractions of carbon atom number from both oils are totally different.

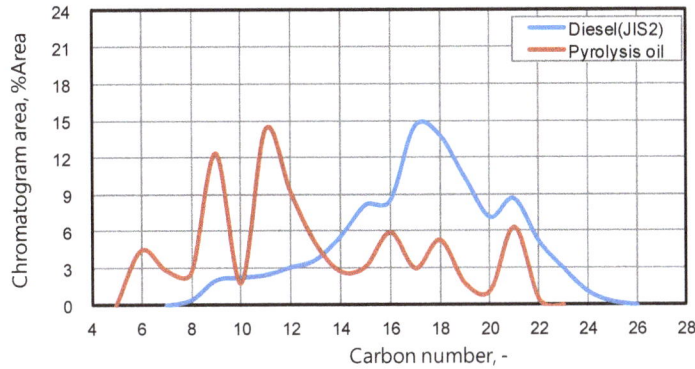

Figure 3. Carbon number distribution

The carbon class of the diesel fuel and the raw pyrolysis oil is illustrated in Figure 4. Comparing to diesel, the content of n-Alkanes which is good for self-ignition in compression ignition engine is very low by around 25%. Aromatic is one of important factors affecting the level of the particulate matter emission from diesel engines (World Wide Fuel Chart, 2006) and is very low. The shape of the carbon class distribution of the raw pyrolysis oil is not like kerosene/jet oil (The American Petroleum Institute Petroleum HPV Testing Group, 2010) and gasoline (Robert A. H. 2004). It mainly consists of iso-alkanes, n-alkanes and olefins in % area of 27%, 25% and 9% respectively. Over 30% content was not able to be defined due to its complicated chemical bond structure. Aromatics, Cyclo-alkanes (Naphthenes), isoalkanes and olefins which are not good for the self-ignition quality (Prasenjeet G., 2006) in compression ignition engine compose around 40%. From this analysis, it can be estimated that the raw pyrolysis oil has very poor self-ignition quality. Thus, the raw pyrolysis oil was mixed with diesel fuel. The mixing ratio was 20, 40, 60 and 80% where the number was the volume percentage of the raw pyrolysis oil to the diesel fuel. In the engine experiment, low mixing ratio blend oil was tried first considering the poor self-ignition quality. By blending, the physical property of the raw pyrolysis oil was improved as shown in Figure 5.

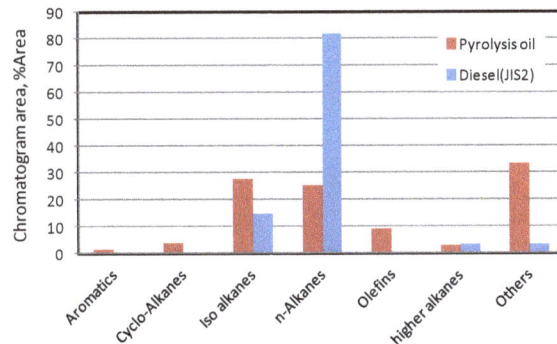

Fugure 4. Hydrocarbon classes

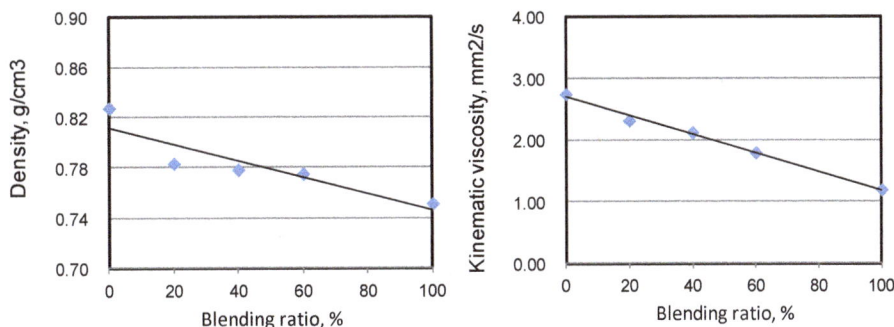

Figure 5. Density and kinematic viscosity with blending ratio (0% : diesel)

2.3 Test Engine

For the engine test, a single cylinder direct fuel injection diesel engine was used and Table 3 shows the engine specification (Koji Y., 2010) (Hiroshi, M., 1999). The engine is very small and turns out a low engine output. The combustion chamber shape was a re-entrant type. The fuel injection timing was 17.5±0.5 degree BTDC CA (Bottom top dead center crank angle) and the injection fuel pressure was 20 MPa, built by the mechanical fuel delivery pump for all tests.

Figure 6 shows the experimental equipment for the engine test. It is consisted of a diesel engine, measurement devices and a dynamometer which controls the engine operation at a certain constant engine load and speed. The in-cylinder pressure data were instantaneously measured by a crystal pressure transducer (KISTLER 6053B). THC and smoke opacity in exhaust gas were measured by the HFID (HORIBA MEXA-1170HFID) and by the opacity type smoke meter (HORIBA MEXA-600S), respectively. CO, CO_2 (Carbon dioxide), NOx and SO_2 (Sulfur oxide) emissions were measured by a NDIR type exhaust gas analyzer (Iwatadengyo Co., Ltd FAST-3100). The intake air temperature and the exhaust gas temperature were measured by a K-type thermocouple with the diameter of 1.0 mm. The fuel consumption was measured by using the electric weighing instruments. All measurement data were continuously processed at the same time in the data collection unit (KEYENCE NR-HA08 and NR-TH08) and a personal computer.

Table 3. Test engine specification

Engine specification	
Cylinder number	1
Bore X Stroke	70mm X 57mm
Displacement volume	219cm3
Compression ratio	20.6
Aspiration type	Natural
Rated power	3kW/3600rpm
Combustion chamber	Re-entrant
Injection pressure	20MPa
Injection timing	17.5±0.5 deg. BTDC
Number of injection hole	4
Diameter of injection hole	0.22mm
Diesel fuel spray angle	95 deg

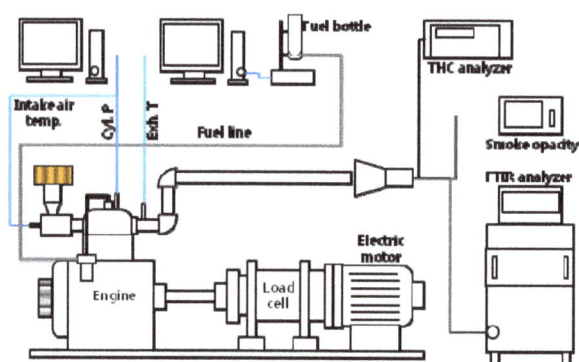

Figure 6. Experimental apparatus

2.4 Test Engine Operation Points

As previously reported, the present paper is aimed to investigate the compatibility of the pyrolysis oil to diesel engine, thus test points in the engine operation were selected based on the US EPA emission legislation test mode shown in Figure 7 (Delphi, 2013).

The eight black-lined circles represent C1-mode which regulates the exhaust emissions for non-road vehicles which operate in construction, air port, mobile cranes, agricultural and forest place. The five blue solid circles are used for the certification test of the exhaust emissions for generation sets. Based on the engine specification shown in Table 3, the engine speed was set to 1500rpm for the idle, 2450rpm for the intermediate and 3500rpm for the rated. The rated engine speed was slightly reduced from its original rated power producing speed, 3600rpm. The 100% power output at the intermediate and the rated engine speed was controlled by conducting the full delivery of the fuel in the injection system for all tested fuels. It means that no additional throttle opening was done to compensate the full load engine output depending on the fuel types. All powers in the partial load operation points were controlled to produce the output powers which were designated to the mode points. When evaluating the experimental results, R25 (25% load at the rated speed) data were interpolated between 50% and 10% because the linearity was shown to be enough.

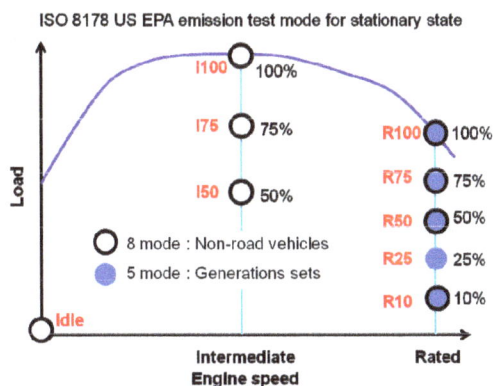

Fugure 7. Test engine operation points

3. Results and Discussion

3.1 Full Load Performance

Full load performance was evaluated upon 100% diesel which is expressed as the blend ratio 0% point on the horizontal axis and the two blend oils, the 20% and 40% blending ratios. For all the full load tests, the throttle position was fully opened as mentioned above. The maximum reachable torque is shown in Figure 8 indicating that it was drastically decreased with the increase of the blend ratio.

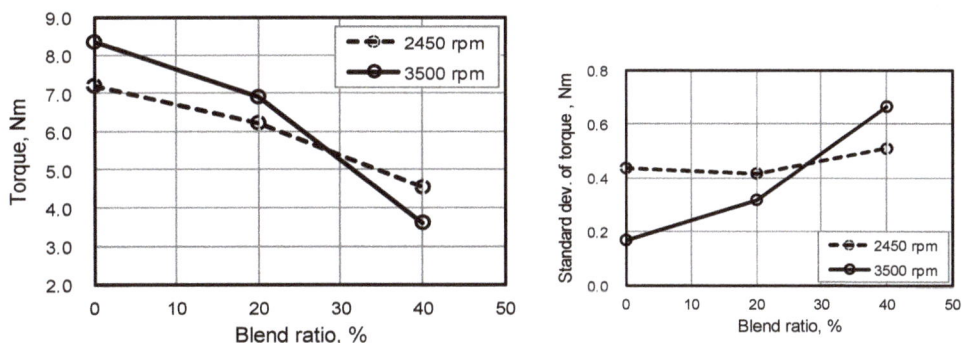

Figure 8. Maximum reachable torque (left) and standard deviation of the torque (right) as a function of the blend ratio (0: 100% diesel)

Up to the 20% blend ratio, comparable engine running was achieved even though the full load output was declined about 13% at 2450rpm and 17% at 3500rpm. The running stability was observed in the same level for the intermediate engine speed and became worse at the rated speed but it was acceptable level because the engine operation was continued without any troubles. For the 40% blend oil, the maximum torque dropped around 37% at the intermediate speed and 56% at the rated speed which corresponds to the R50 output of diesel. The engine operation stability which is expressed by the standard deviation of the torque depicted in Figure 8 (right one) became worse with increasing the blend ratio and in the case of the 40% blend oil at 3500rpm, the engine could keep running only for a few minutes.

The heat value of the raw pyrolysis oil is almost identical with the diesel fuel and this means there is no change in the heat value by blending the pyrolysis oil with diesel fuel. Nevertheless, the torque of those blend oils dropped and some analysis is followed.

For the case of I100 operation point, the injection quantity decreased 5.8mg/st (-8%) for the blend 20% and 4.6mg/st (-27%) for the blend 40% comparing to 6.3mg/st of diesel fuel with increasing the blend ratio as shown in Figure 9. It is thought that the physical property of the raw oil such as lower density and kinematic viscosity affected the injection amount (Desantes, J., M., 2003) (Qaisar, H., 2013). The density drops, 5.3% for the blend 20% and 5.9% for the blend 40% as shown in Figure 4, can not cover all the injection amount drops. In the event of the injection, the rate of injection might be additionally reduced by the lower kinematic viscosity (Chang, C. T., 1997). The relationship between the injected amount to the torque can be simply calculated by dividing the engine torque by the injection quantity. The fuel mass to torque conversion factor is around 1.1 Nm/[mg/st] for the diesel shown in Figure 9. For the 20% blending oil, 6.8Nm can be estimated by using the conversion factor and the injection mass of 5.8mg/st which was fully delivered by the pump but the actual torque gained was 6.2Nm (8% lower). For the 40% blend oil, the calculated torque is 5.2Nm and the actual was 4.5Nm (27% lower).

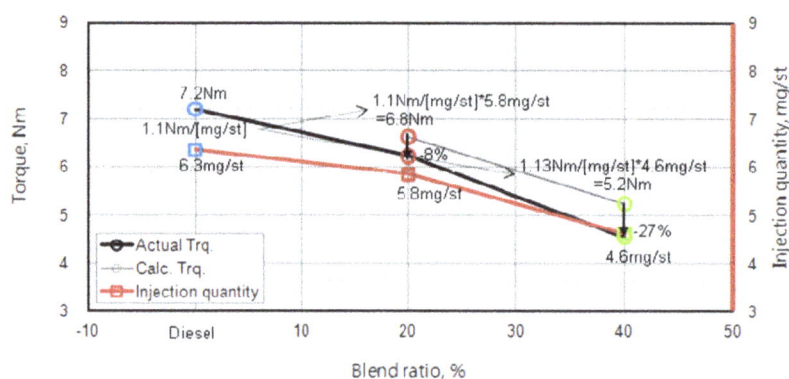

Figure 9. Torque and injection quantity as a function of the blend ratio at I100, 2450rpm

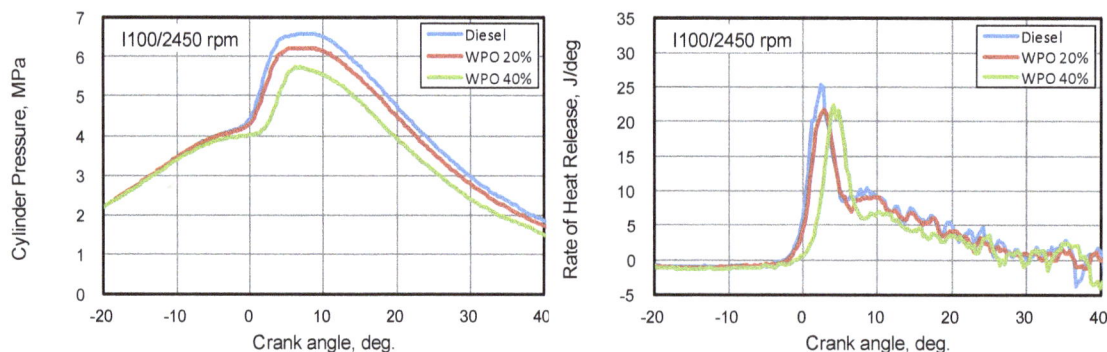

Figure 10. Cylinder pressure and rate of heat release at I100, 2450rpm

For both blend oils, the actual torque output was less than the estimated ones and this went worse with increasing the blend ratio. This can be explained based on the combustion analysis as depicted in Figure 10. In the premixed combustion (John, B., H., 1988) (Johann, H., 2013) (Horn, U., 2007) where the fuel-air mixture which was accumulated during the ignition delay defined as duration between the injection timing shown in Table 3 and the first rise of the heat release rate (Carroll E. G., 1998) (Ghojel, J., 2005) rapidly exploded, the cylinder pressure rise was slower due to the lower injection rate caused by the physical oil property as described above. The ignition delay became longer with increasing the blending ratio. The enlarged ignition delay is caused by the chemical property of the raw pyrolysis oil based on the analysis result shown in Figure 4. For the 20% blending ratio, even though the ignition timing was slightly delayed, the amount of the heat release in the premixed combustion phase was lower than for the diesel and the phase was slightly shifted backward. As a result, the torque dropped more than the estimated one. This might be caused by mainly the lower injection rate due to the physical property of the raw pyrolysis oil. For the 40% blending ratio, the ignition delay was significantly enlarged and the premixed heat release phase was moved more than that of the 20% blend. Thus the larger torque drop than the calculated one was caused mainly by the ignition delay due to the chemical property of the raw pyrolysis oil.

The identical analysis can be basically propagated for the case of R100 operation point. The injection quantity of the diesel fuel increased more than that of I100 due to the increased pressure build-up driven by the higher engine speed and the injection quantity of 6.9mg/st produced the torque of 8.3Nm and the torque conversion factor was 1.2Nm/[mg/st] as illustrated in Figure 11. The estimated torque was 7.6Nm for the 20% blending and 5.4Nm for the 40% blending. The actual torque deteriorated by -9% and -35%, respectively. Those numbers are even larger especially for the 40% blend than those of the I100. For the 20% blending, the amount of the heat release in the premixed combustion phase decreased slightly than that of I100 shown in Figure 12. For the 40% blend, the heat release phase delayed more by greatly enlarged ignition timing due to the higher piston speed and this led the magnitude of the actual torque drop to the calculation to be larger than the case of I100. The ignition delay will be summarized later including part loads.

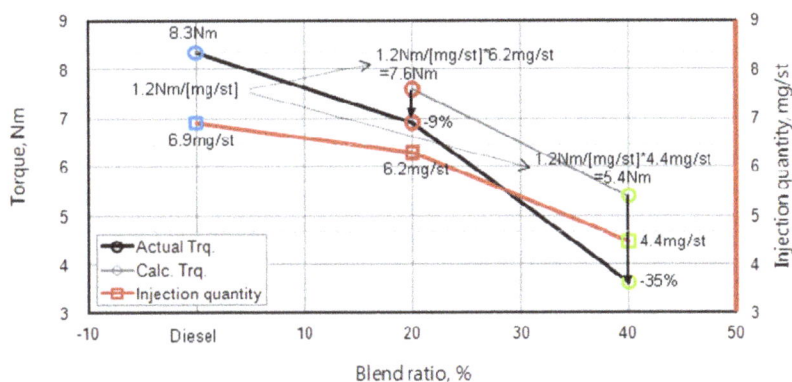

Figure 11. Torque and injection quantity as a function of the blend ratio at R100, 3500rpm

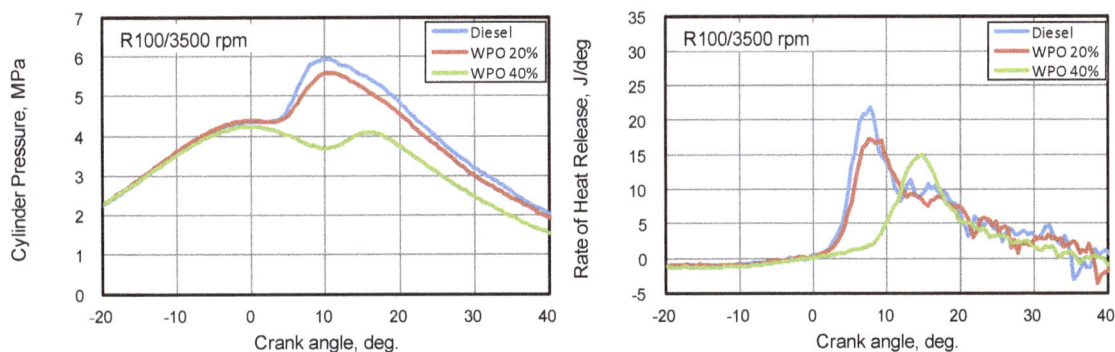

Figure 12. Cylinder pressure and rate of heat release at R100, 3500rpm

3.2 Exhaust Emissions

The concentration based exhaust emissions in the experiment was processed to weight based ones using the carbon balance and the fuel consumption so that the compatibility of the pyrolysis oil to diesel can be evaluated according to the US EPA regulation. The US EPA defines the Tier1 regulation as 10g/kWh of NOx+NMHC (Non Methane HydroCarbon), 8g/kWh of CO and 1g/kWh of PM (Partuculate matters) for the test engine category (EPA) using the test mode described in the section 2.4. The regulation emission numbers were used as just reference value in the report. For the PM, the smoke opacity was converted to the mass value (Diesel-net) only for the R100 operation point.

Figure 13 shows the result of the exhaust emissions for the intermediate engine speed, 2450rpm. The NOx emission was not changed in overall by the 20% blending and tended to decrease for the 40% blending. For the

20% blending ratio, even with the enlarged ignition delay which is shown in Figure 14 and 15, the fraction of premixed combustion was slightly lower than that of diesel and this was caused by the lower injection rate as described previously. For the 40% blending, the ignition timing delayed more and the premixed combustion fraction increased more than the 20% blending but the combustion occurred in more expanded space according to the downward of the piston and this resulted in the less NOx emission even in small. The CO emission decreased for the 20% blending and increased again for the 40% blend. This tendency almost tracked with the fraction of the premixed combustion excepting the I100. The more the premixed combustion was, the higher the CO emission was, which is summarized in Figure 16. There was no significant change in the THC emission.

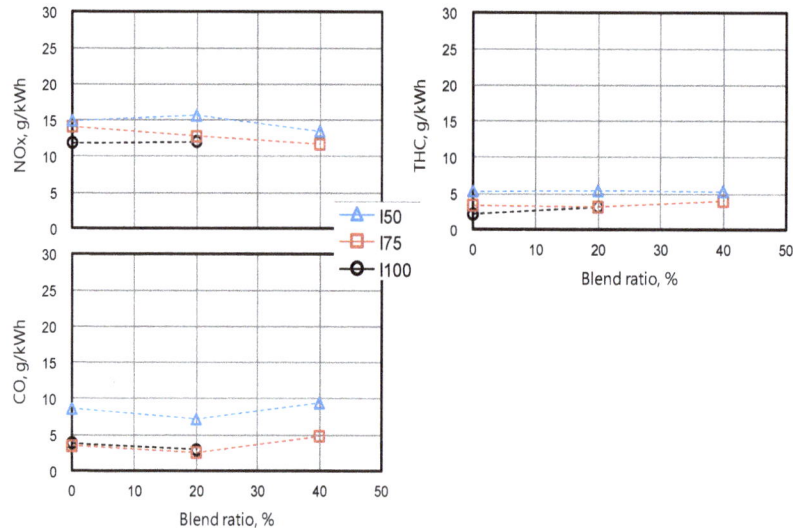

Figure 13. NOx, THC and CO emissions as a function of the blend ratio at the intermediate engine speed, 2450rpm (0: 100% diesel)

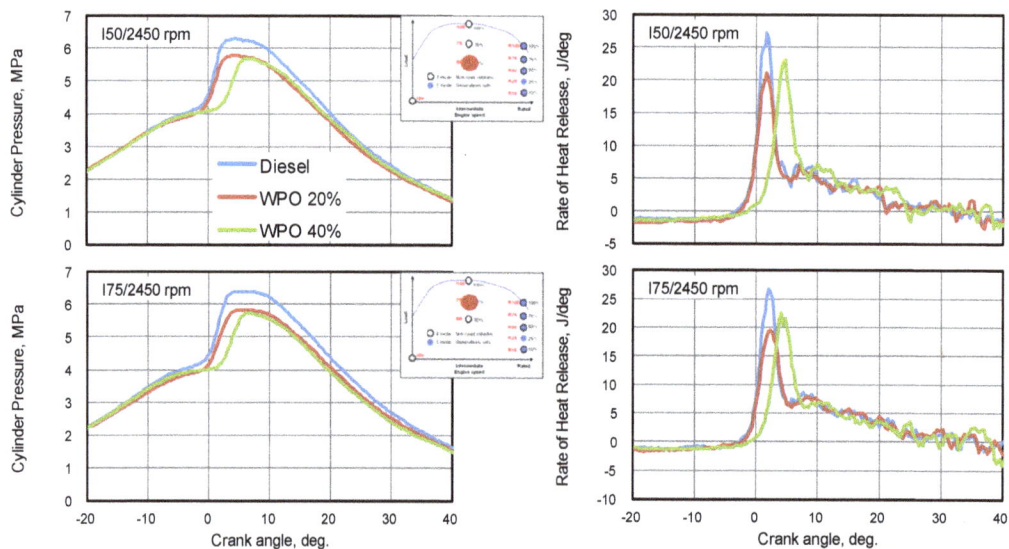

Figure 14. Cylinder pressure and rate of heat release at the intermediate engine speed, 2450rpm

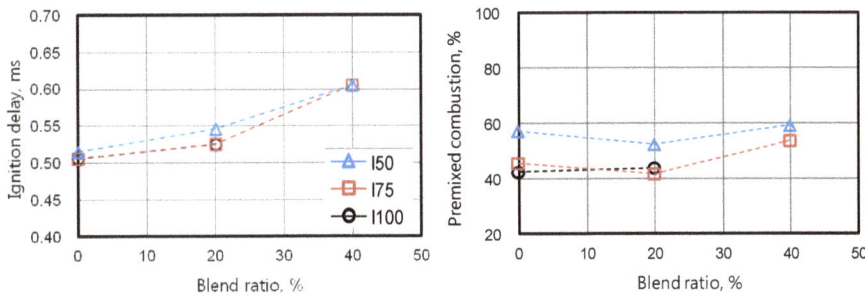

Figure 15. Ignition delay and premixed combustion fraction as a function of the blend ratio at the intermediate engine speed, 2450rpm (0: 100% diesel)

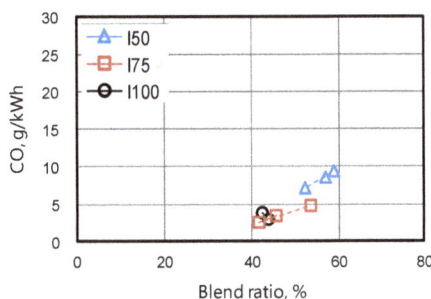

Figure 16. The effect of the premixed combustion fraction on the CO emission at the intermediate engine speed, 2450rpm (0%: diesel)

At the rated engine speed, the NOx emission was almost identical tendency with the result of the intermediate engine speed but decreased around double in the 40% blending for R10 (Figure 17) due to the much longer ignition delay as shown in Figure 18 and 19. The THC and CO emissions increased with the increase of the blending ratio at a very low load such as R10. For the 40% blending, the ignition delay was not different from those of the intermediate speed but the ignition timing was too much retarded according to the increase of the engine speed to keep the stable engine operation for the R10 and R50. The PM emission for the 20% blending was slightly higher but the absolute number was far less the EPA regulation, 1g/kWh. In Figure 20, the impact of the premixed combustion fraction on the CO emission is depicted and almost the same tendency as seen in I100 is found

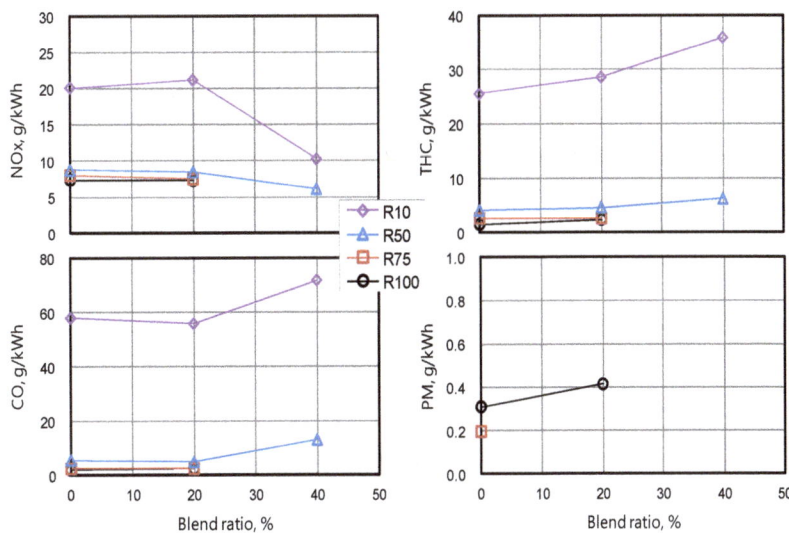

Figure 17. NOx, THC and CO emissions as a function of the blend ratio at the rated engine speed, 3500rpm. (0%: diesel)

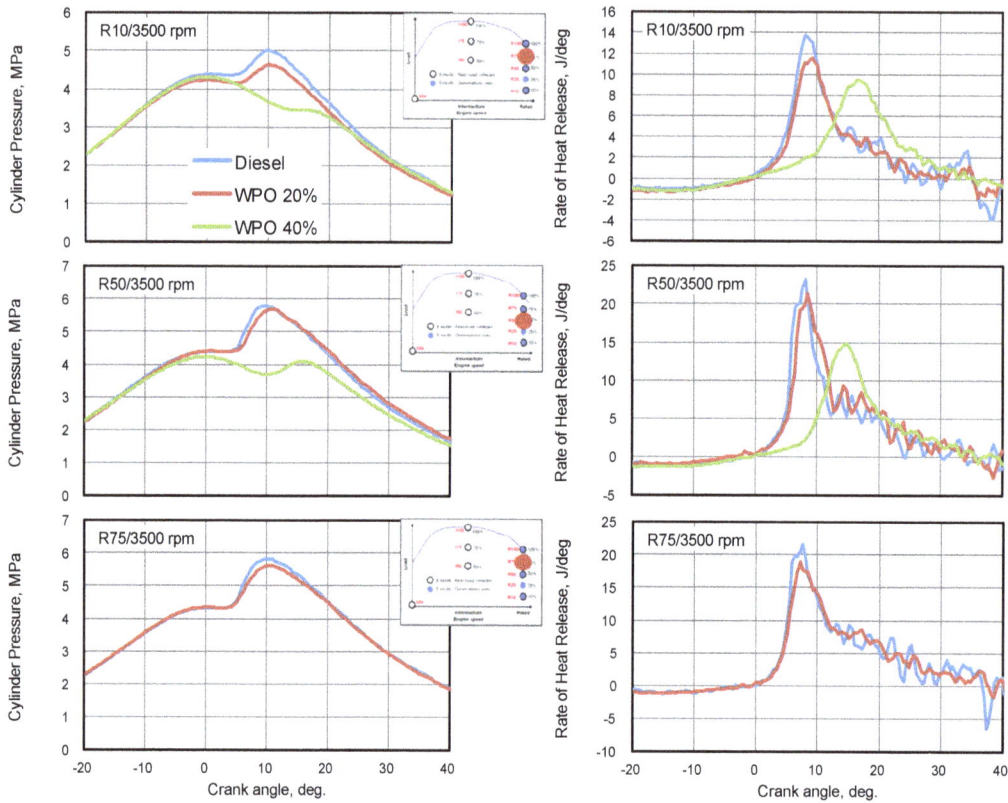

Figure 18. Cylinder pressure and rate of heat release as a function of the blend ratio at the rated engine speed, 3500rpm

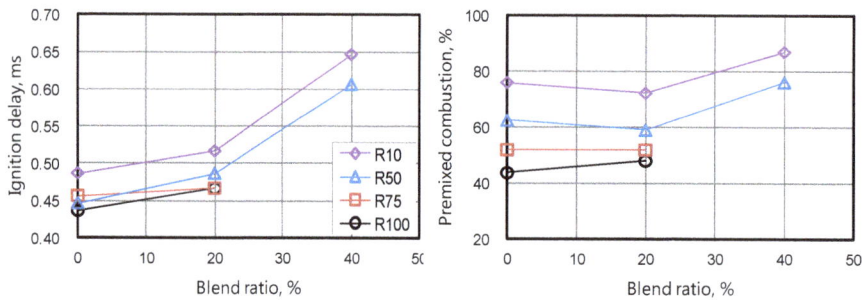

Figure 19. Ignition delay and premixed combustion fraction as a function of the blend ratio at the intermediate engine speed, 3500rpm (0%: diesel)

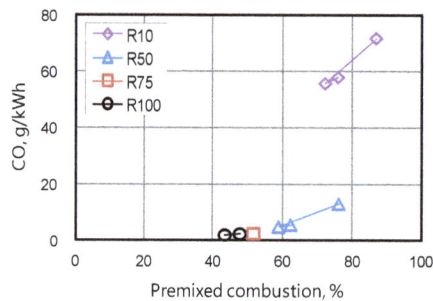

Figure 20. The effect of the premixed combustion fraction on the CO emission at the intermediate engine speed, 2500rpm (0%: diesel)

Figure 21 shows the exhaust emission with regard to C1-8 mode and D2-5 mode. For the case of diesel fuel, the NOx+THC emission was slightly over the Tier 1 regulation. The test engine was not new one and had been run with various non regulated fuels such as biomass oils and other pyrolysis oils. All emission measurement devices were not fully pre-conditioned for the engine test like an emission certification test. Considering such conditions, the NOx+THC emission result can be acceptable. The THC emission is actually slightly higher than NMHC which is defined as the regulation but the amount of the deviation was low enough to be neglected. The CO emission was lower than the regulation. The 20% blending oil (blend20) showed the same emission level for the NOx+THC emission and a lower emission level for the CO emission. This tendency is the same for the both test modes, the C1-8 mode and the D2-5 mode.

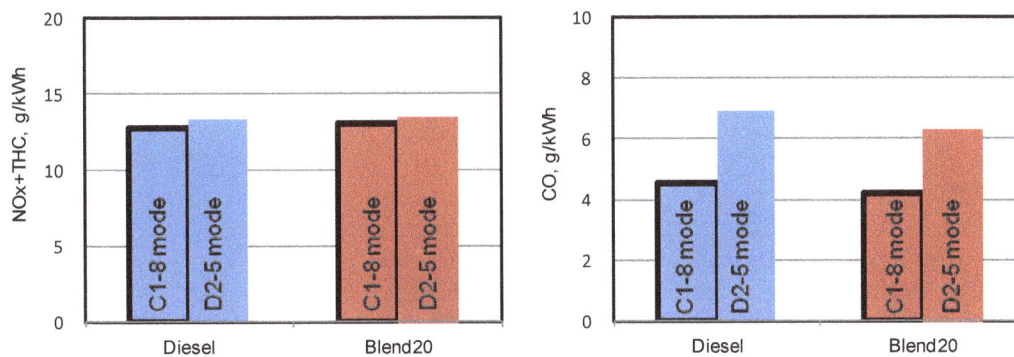

Figure 21. NOx+THC and CO emissions in the C1-8 mode and D2-5 mode for diesel and 20% blending

Many new pyrolysis oils derived from wastes such as plastics and biomass have various kinds of undefined exhaust emissions in terms of the amount and the type from its combustion inside engines. In this report, toluene, acetaldehyde and formaldehyde were defined as toxic hydrocarbon matters. The toxic hydrocarbon ratio was expressed as the ratio of the three toxic hydrocarbons content in the total hydrocarbon emission and is compared in Figure 22. The toxic hydrocarbon ratio was slightly lower for the 20% blending than for the diesel fuel in all engine operation points.

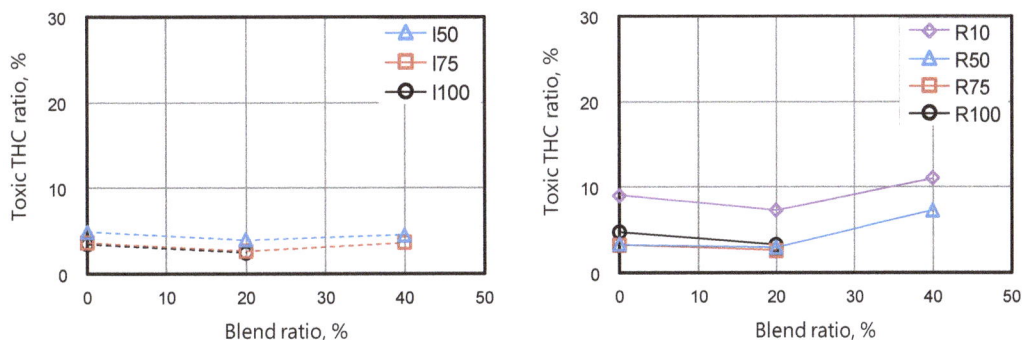

Figure 22. Comparison of the toxic hydrocarbon ratio (0%: diesel)

3.3 Thermal Efficiency

The brake thermal efficiency at full load operation points is plotted in Figure 23. The thermal efficiency decreased with the increase of the blend ratio because of the less injection rate due to the physical property of 20% blend and the retarded ignition timing caused by the chemical property of 40% blend. The magnitude of the deterioration was slightly higher at 3500rpm than at 2450rpm due to the increased piston speed. For the 40% blend, at 3500rpm, the engine operation was too unstable to keep the operation for prolonged time but at 2450rpm, the engine could run normally. Thus, in order to increase the useful amount of the pyrolysis oil, 40% blend at 2450rpm as shown in the right one of Figure 23 can be one choice for the electricity generation along with the de-rated engine power by one third and the lower thermal efficiency and the acceptable exhaust emission.

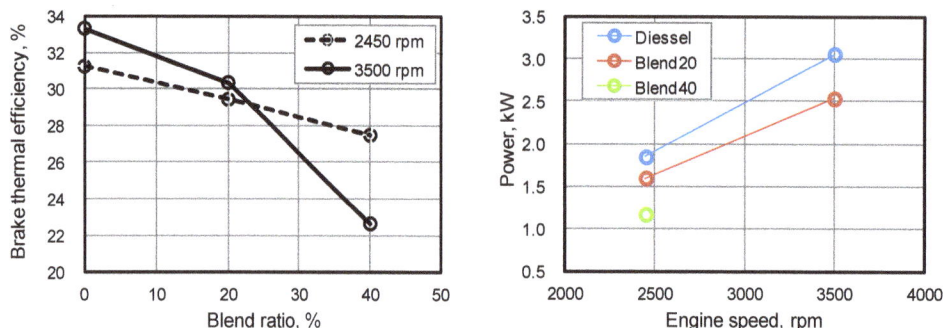

Figure 23. Brake thermal efficiency as a function of the blend ratio (0%: diesel) and possible engine operation points for the electricity generation (Right)

4. Conclusion

The pyrolysis oil was produced from real household waste plastics in a commercial pyrolysis plant. The raw pyrolysis oil was blended with diesel fuel in 20% and 40% volumetric ratio because of its low kinematic viscosity and poor self-ignition quality. A small single cylinder air cooled direct injection diesel engine was used for the combustion experiment to compare with the operation by diesel fuel.

With 20% blending ratio, the compatible engine operation was observed with regard to the full load engine output, the exhaust emissions and the thermal efficiency. The maximum engine power dropped around 13% at the intermediate engine speed and 17% at the rated engine speed. The exhaust emissions during the operation under the US EPA standard test modes, the C1-8 mode for non-road vehicles and the D2-5 mode for generation sets, resulted in the same level for the NOx+THC emission and lower for the CO emission and slightly higher for the PM just at the rated power point but the number was much less than the regulation. The toxic hydrocarbon which was defined as the content of toluene, formaldehyde and acetaldehyde in the THC emission was as low as diesel. The thermal efficiency at full load was lower by maximum 3% in absolute. The combustion was mainly characterized by the oil physical properties such as density and kinematic viscosity and the impact of the chemical property was minor.

For 40% blending ratio, the maximum engine power dropped 37% at the intermediate engine speed and 57% at the rated engine speed where the engine operation was possible to be kept just for a few minutes. If the de-rated power by around one third is acceptable, the engine operation at 2450rpm is can be one choice for the sake of decreasing the blend fraction of the diesel fuel. The chemical property played major role in the combustion characterized with the long ignition delay.

So the present pyrolysis oil can be used in a diesel engine with around 20% blending without any limitations and with 40% blending at a limited engine speed such as below 2450rpm.

Acknowledgments

The authors would sincerely like to thank for the cooperative support for this study provided by Mr. Suzuki, Mr. Yamazawa and other students in the Yoshida Internal Combustion Engine Laboratory at Nihon University

References

American Society for Testing and Materials, ASTM D975: Standard Specification for Diesel Fuel Oils.

Carroll, E. G. (1998). Engine Heat Release via Spread Sheet. *American Society of Agricultural Engineer, 41*(5), 1249-1253. http://dx.doi.org/10.13031/2013.17290

Chang, C. T., & Farrell, P. V. (1997). A study on the effects of fuel viscosity and nozzle geometry on high injection pressure diesel spray characteristics. No. 970353. *SAE Technical Paper,* 1997.

Daniel, H., & Perinaz, B. (2012). WHAT A WASTE A Global Review of Solid Waste Management, Urban Development & Local Government Unit, The World Bank, March 2012, No. 15.

Desantes, J. M., et al. (2003). Measurements of spray momentum for the study of cavitation in diesel injection nozzles. No. 2003-01-0703. *SAE Technical Paper,* 2003.

Ghojel, J., & Damon, H. (2005). Heat release model for the combustion of diesel oil emulsions in DI diesel engines. *Applied Thermal Engineering, 25*(14), 2072-2085. http://dx.doi.org/10.1016/j.applthermaleng.2005.01.016

Hiroshi, M., Kozaburo, W., Koji, Y., Hideo, S., & Hidenori, T. (1999). Combustion Characteristics and Exhaust Gas Emissions of Lean Mixture Ignited by Direct Diesel Fuel Injection with Internal EGR. *The Engineering Society For Advancing Mobility Land Sea Air and Space*, Small Engine Technology Conference and Exposition Madison, Wisconsin, Sep. 28-30, 1999, SAE 1999-01-3265.

Horn, U., Egnell, R., & Johansson, B. (2007). Detailed heat release analyses with regard to combustion of RME and oxygenated fuels in an HSDI diesel engine. *SAE Technical Paper*, 2007, No. 2007-01-0627. Retrieved from http://delphi.com/pdf/emissions/Delphi-Heavy-Duty-Emissions-Brochure-2013-2014.pdf

Jane, P., & Krongkaew, L. (2013). Performance and Emission Evaluation of Blends of Diesel fuel with Waste Plastic Oil in a Diesel Engine. *International Journal of Engineering and Innovative Technology (IJESIT), 2*(2), March 2013, ISSN: 2319-5967

Johann, H., & Kar, H. (2013). ENGINE-BASED TEST METHOD FOR DETERMINING THE IGNITION QUALITY OF DIESEL FUELS. *MTZ, 74*.

John, B. H. (1988). Internal Combustion Engine Fundamentals. McGraw-Hill Book Company, 503-506, ISBN 0-07-100499-8.

Kaustav, M. C. T. (2014). Performance and Emission Test of Several Blends of Waste Plastic Oil with Diesel and Ethanol on Four Stroke Twin Cylinder Diesel Engine. *IOSR Journal of Mechanical and Civil Engineering (IOSR-JMCE), 11*(2 Ver. I), 47-51. E-ISSN: 2278-1648, p-ISSN:2320-334X

Koji, Y. (2010). Application of Cellulosic Liquefaction Fuel (CLF) and Fatty Acid Methyl Ester (FAME) Blends for Diesel Engine. *SAE International Journal of Fuels and Lubricants, 3*, 1093-1102.

Lacey, P. I. (1992). THE RELATIONSHIP BETWEEN FUEL LUBRICITY AND DIESEL AND DIESEL INJECTION SYSTEM WEAR, Southwest Research Institute, Interim Report BFLRF, No. 275, Mar. 27, 1992.

Luca, L. R. L., Gianmarco, P., & Mohamad, T. (2013). World Energy Resources: Waste to Energy, *World Energy Council*.

Nilamkumar, S. P., & Kehur, D. D. (2013). Waste Plastic Oil As a Diesel Fuel In The Diesel Engine:A Review. *International Journal of Engineering Research & Technology (IJERT), 2*(3). March-2013, ISSN: 2278-0181.

Pawar, H. R., & Lawankar, S. M. (2013). Waste plastic Pyrolysis oil Alternative Fuel for CI Engine – A Review. *Research Journal of Engineering Sciences, 2*(2), 26-30, February (2013) ISSN 2278-9472.

Prasenjeet, G., Karlton, J. H., & Stephen, B. J. (2006). Development of a Detailed Gasoline Composition-Based Octane Model. *Ind. Eng. Chem. Res., 45*, 337-345. http://dx.doi.org/10.1021/ie050811h

Qaisar, H., et al. (2013). Mathematical Modeling of Fuel Pressure inside High Pressure Fuel Pipeline of Combination Electronic Unit Pump Fuel Injection System. *Research Journal of Applied Sciences, Engineering and Technology, 6*(14), 2568-2573. ISSN: 2040-7459; e-ISSN: 2040-7467.

Rajesh, G., Deva, K., & Vijaya, K. R. (2011) EXPERIMENTAL EVALUATION OF A DIESEL ENGINE WITH BLENDS OF DIESEL-PLASTIC PYROLYSIS OIL. *Internal Journal of Engineering Science and Technology (JIEST), 3*(6), ISSN: 0975-5462.

Robert, A. H., & Andrew, J. K. (2004). Chemical Composition of Vehicle-Related Volatile Organic Compound Emissions in Central California. Final Report Contract 00-14CCOS.

Senthilkumar, T., & Chandrasekar, M. (2012). The Evaluation of Blend of Waste Plastic Oil-Diesel fuel for use as alternated fuel for transportation. *2nd International Conference on Chemical, Ecology and Environmental Sciences (ICCEES2012) Singapore April 28-29*.

Stefano, F., Roberto, G., Maurizia, S., & Monica, P. (2013). Diesel Fuel by Scrap-Tyre Thermal-Mechanical Pyrolysis. *SAE International*.

The American Petroleum Institute Petroleum HPV Testing Group. (2010). KEROSENE/JET FUEL CATEGORY ASSESSMENT DOCUMENT, Consortium Registration

United Nations. (2013). Department of Economic and Social Affairs, Population Division.World Population Prospects: The 2012 Revision, DVD Edition.

World Energy Council. (2013). World Energy Scenarios, Composing energy futures to 2050, Project Partner Paul Scherrer Institute (PSI), Switzerland, World Energy Council 2013.

World Wide Fuel Chart. (2006). Forth Edition.

10

Investigation of Air and Air-Steam Gasification of High Carbon Wood Ash in a Fluidized Bed Reactor

Adrian K. James[1], Steve S. Helle[2], Ronald W. Thring[2], P. Michael Rutherford[2] & Mohammad S. Masnadi[3]

[1] Natural Resource and Environmental Science, University of Northern British Columbia, Prince George, BC, Canada

[2] Environmental Science & Engineering, University of Northern British Columbia, Prince George, BC, Canada

[3] Chemical and Biological Engineering, University of British Columbia, Vancouver, BC, Canada

Correspondence: Ronald W. Thring, Environmental Science & Engineering, University of Northern British Columbia, Prince George, BC, Canada. E-mail: thring@unbc.ca

Abstract

The pulp and paper industry in an effort to offset fossil fuel demand uses woody biomass combustion as a renewable energy source to meet their ever-growing energy demands. Boiler combustion systems are often used to provide this energy. However, large amounts of high carbon ash are produced from some boilers resulting in technological, economic and environmental challenge. This high carbon ash is considered to be of very little economic and environmental value and is typically sent to landfills. Reuse of this ash in some boilers requires upgrading and is not economically feasible. Therefore, this study investigates the feasibility of gasifying high carbon wood ash of particle sizes smaller than 3 mm, while comparing its behaviour to that of unburned wood. Gasification was conducted in a stainless steel bubbling fluidized bed reactor 3-inch diameter and height of approximately 800 mm using air and air-steam as gasifying agents. Parameters of interest included equivalence ratio (ER), gas calorific value, carbon conversion efficiency and produced gas yield. High carbon ash was successfully gasified at low temperatures and atmospheric pressure and showed similar trends as woody biomass. The higher heating value (HHV) and carbon conversion efficiency increased with increasing temperature. The H_2/CO molar ratio was higher for the air-steam process. Future areas of research could include investigating the viability of producing a gas of even higher heating value.

Keywords: gasification, fluidized bed reactor, high carbon wood ash, air/air-steam gasification, higher heating value, carbon conversion

1. Introduction

The pulp and paper industry produces large volumes of high carbon ash from boilers. While the carbon content is relatively high in this residue, the energy content of all of the ash produced is approximately 1% of the energy content of the wood. Combustion of the high carbon ash presents a number of operational problems such as corrosion and scouring. Due to the design of some boilers the ash may be carried by the flue gas through the boiler tubes creating problems. The variation in particle sizes also causes inherent problems in a fixed bed system. Fixed bed systems usually require a uniform feedstock to avoid channelling (Warnecke, 2000; Ryu et al., 2006). Exorbitant costs are also associated with design and system alterations.

High carbon ash is considered to have very little potential economic and environmental benefit at this stage and is typically sent to landfills. The application of bottom ash as a soil additive is restricted in British Columbia, Canada. Other options for utilizing high carbon ash must be explored, including use as a low cost feedstock for already existing gasifiers in order to recover as much energy as possible, while reducing ash volume.

Biomass gasification is a thermo-chemical process of gaseous fuel production by partial oxidation of a solid fuel to produce heat, electricity and synthesis gas (Rade & Karamarkovic, 2010). Gasification results in producer gas containing CO, H_2, CnHx and other gases (Turare, 1997). The main objective is to generate a combustible gas rich in CO, H_2 and CH_4 with a medium to high lower heating value (LHV) (Alauddin, Lahijani, Mohammadi, & Mohamed, 2010; Skoulou, Koufodimos, Samaras, & Zabanioutou, 2008). Operating conditions such as temperature, equivalence ratio (ER) and steam/biomass (S/B) ratio play important roles in biomass gasification.

Bed temperature is one of the most important operating parameters in gasification, affecting both the heating value and the producer gas composition (Alauddin et al., 2010). The heat needed for air gasification is provided by partial combustion of the biomass. High temperatures increase CO_2 production, lowering the heating value of the produced gas. A high bed temperature improves carbon conversion and steam cracking and reforming of tars, resulting in less char, reduced tar formation and higher gas yields (Alauddin et al., 2010; Pinto et al., 2003; Chairprast & Vitidsant, 2009).

The equivalence ration (ER) also strongly influences the gasification product composition. More combustion occurs at high ER, increasing CO_2 production (Alauddin et al., 2010; Mandl, Obernberger, & Biedermann, 2010). A higher air flow rate results in higher gas velocities, improving the combustion of solid char due to improved oxygen mass transfer (Mandl et al., 2010; Natarajan, Nordin, & Rao, 1998). An equivalence ratio of 0.2–0.3 is most favourable for producing CO-rich gas (Li et al., 2004). When steam is the gasifying agent, H_2 and CO_2 increase, while CO decreases due to the water gas shift reaction (Devi, Ptasinski, & Janssen, 2003).

Fluidized bed reactors have been widely applied for gasification, pyrolysis and combustion of a wide range of particulate materials including biomass (Cui & Grace, 2007). Advantages include high heat transfer, uniform and controllable temperatures, favourable gas-solid contact and the ability to handle a wide range of particulate properties such as particle diameter. Fluidized bed reactors also accommodate wide variations in fuel quality.

Air-blown biomass gasification produces low calorific value gases, with higher heating values (HHV) of 4-7 MJ/Nm^3, whereas oxygen and steam-blown processes result in a HHV of 10–18 MJ/Nm^3 (Li et al., 2004). Circulating fluidized bed (CFB) tests using various feedstocks such as, spruce-pine-fir sawdust mixture, 1:1 ratio of pine bark and spruce whitewood mix, cypress, hemlock and cedar-hemlock mixtures have produced gases with HHV from 2.43–6.13 MJ/kg, with either air or air-steam as the gasifying agent. For example, in a fluidized bed experiment at atmospheric temperature, the gasification of pine sawdust produced a LHV of 6.74–9.14 MJ/Nm^3 in an air-steam medium at ER = 0.22 (Lv et al., 2004).

The research carried out in this study was intended to determine the feasibility of gasifying high carbon wood ash particles smaller than 3 mm to identify whether they behave similarly to unburned wood when gasified. Test were carried out to,

(1) Determine the range of equivalence ratios for stable operation.

(2) Determine the calorific value of the producer gas with air and air-steam agents to ascertain the potential of producing a low to medium calorific value syngas.

(3) Measure the carbon conversion efficiency.

(4) Calculate the product gas yield.

2. Experimental

2.1 Feed Materials

Wood ash particles from from an industrial scale fixed-bed boiler (Canfor Pulp Mill, Prince George, BC) constituted the feedstock in study. Hog fuel is used in this boiler, comprised predominantly of softwood sawmill waste derived from pine wood. Silica sand was the inert bed material. The proximate and ultimate analyses of the ash are provided in Table 1.

Table 1. Proximate and ultimate analyses of hog fuel

Higher heating value (MJ/kg)	11.60
Proximate analysis (wt.% dry basis)	
Volatile Matter	21.5
Fixed Carbon	28.8
Ash	49.7
Ultimate analysis (wt. % dry basis)	
C	48.5
H	0.9

O	33.4
N	0.2
S	0
Other ash forming elements	17.0

O – Calculated by difference.

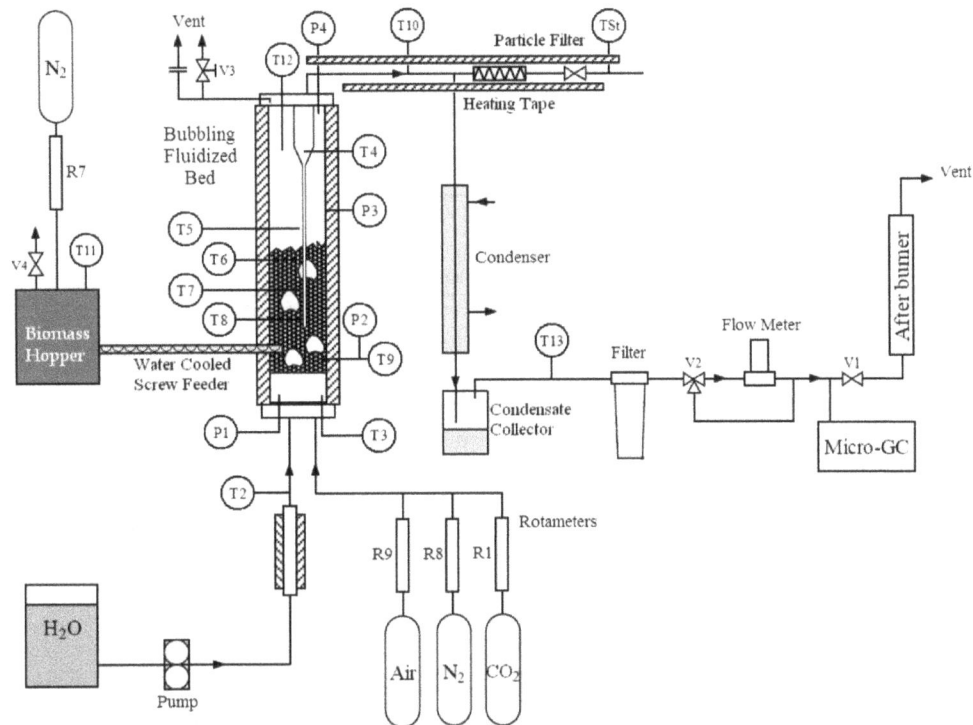

Figure 1. Schematic diagram of experimental unit for biomass air and air-steam gasification in a bubbling fluidized bed. T – Thermocouple P- Pressure sensor R - Rotameter

2.2 Gasificaion Setup

Air gasification and air-steam gasification were carried out in the lab-scale bubbling fluidized bed reactor shown schematically in Figure 1. Constructed from 310 stainless steel 3-inch diameter (nominal) pipes (I.D. = 77.9 mm) of height of approximately 800 mm. Two electrical heaters supplied heat to the reactor. The reactor was charged with 1.4 kg of sand as bed material. A pressure tap located in the biomass feeder was used to control and facilitate the discharge of feedstock. The bed was fluidized by air and nitrogen introduced below the distributor. Water was pumped to the reactor, then vapourized, with its flow rate measured by a steam flow meter. The biomass feedstock was fed from the side of the reactor through an atomizer nozzle, covered by a cooling jacket to keep the feedstock temperature below 80 °C, to avoid plugging by thermal decomposition. The product gas left the reactor at the top and passed through a cyclone to return solid to the gasification bed. Excess steam in the product gas was separated by a condenser, while fine ash and char particles were captured by an internal cyclone, supplemented by a filter after the condenser and a waste bin. The product gas flow rate was measured by a rotameter combined with a thermocouple and a pressure transducer.

2.3 Experimental Procedure

The feedstock was added to the hopper prior to each experimental run. The gasifier and furnace heaters for air preheating were turned on, and controllers were set at the selected operating temperatures. With sand as the bed material, the reactor was charged with ~ 7 L/min of nitrogen to assist with fluidization and aid heat transfer. The feedstock was then fed at 176 g/h, with an air supply of 0.282 Nm³/h from the bottom of the reactor to provide an ER of 0.12. When the system reached steady state, gas samples were taken at 4 min intervals. Experiments were conducted at various bed temperatures within the range of 650–770 °C. The reactor was then operated at a fixed temperature of 775 °C while varying the ER.

For air-steam gasification, water was introduced to the reactor to provide different steam/biomass (S/B) ratios at a fixed temperature of 715 °C and a fixed ER of 0.12, with gas sampling as for air alone.

2.4 Gas Analysis

The concentrations of H_2, N_2, CO, CO_2 and C_nH_x were measured by a micro-gas chromatograph CP-4900 (Varian Inc.) equipped with a CO_x column and a thermal conductivity detector.

2.5 Analyses of Experimental Results

To assess the gasification process, variables such as Equivalence Ratio (ER), carbon conversion efficiency and higher heating values (HHV) were determined as follows.

$$ER = \frac{mas\ flow\ of\ air\ /mass\ flow\ dry\ biomass}{stoichiometric\ air/biomass\ ratio} \tag{1}$$

$$S/B = \frac{steam\ feed\ rate + moisture\ introduced\ with\ fuel\ (g/h)}{Total\ fuel\ feed\ rate\ (g/h)} \tag{2}$$

$$Carbon\ conversion\ efficiency = \frac{gas\ velocity\ x1000[CO\%+CO2\%+3(C3H8\ \%)]\frac{12}{24.79}}{biomass\ feed\ flow\ rate\ x\ C\%} \tag{3}$$

where produced gases are in volume%, gas flow rate is (Nm^3/h), feed flow rate (g/h) and C%, is the biomass percent carbon based on the ultimate analysis. The higher heating value is estimated from

$$HHV = (12.75\ H_2 + 12.63\ CO + 39.82\ CH_4 + 63.43\ C_2H_4 + 99\ C_3H_8...)/100 \tag{4}$$

where the species contents are in mol% and their heats of combustion are in $MJ/Nm3$ (Li et al., 2004; Zhang, 2011).

3. Results

3.1 Air Gasification

3.1.1 Effect of Reactor Temperature

Analysis of the gas produced was carried out for reactor temperatures ranging from ~ 650 to ~ 770 °C in increments of approximately 30 °C. From Figure 2, it can be seen that the CO concentration increased with temperature. All other gas concentrations remained nearly constant, except for CO_2 whose concentration decreased with increasing temperature. The gases produced were predominantly influenced by the reactions:

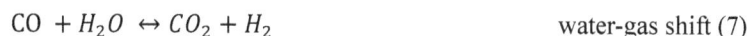

$$C + \tfrac{1}{2}O_2 \rightarrow CO \qquad\qquad\qquad \text{partial oxidation} \tag{5}$$

$$C + O_2 \rightarrow CO_2 \qquad\qquad\qquad \text{complete combustion} \tag{6}$$

$$CO + H_2O \leftrightarrow CO_2 + H_2 \qquad\qquad\qquad \text{water-gas shift} \tag{7}$$

Hence the C present in the fuel as char reacted directly with the O supplied by the air to produce CO, an exothermic reaction. CO production favoured higher temperatures, resulting in less CO_2 generation with increasing temperature. The reactions were being carried out at ER = 0.12, below the ideal ER range of 0.2–0.3 (Li et al., 2004). The limited O_2 fed resulted in a high $CO:CO_2$ ratio. This would result in greater concentrations of CO instead of CO_2. The H_2 concentration remained low and relatively constant, in part because there was very little H in the fuel. CH_4 could not be detected but propane was found at very low concentrations. Table 2 summarizes the results when temperature was varied during air gasification. The carbon conversion efficiency increased from 31.2 to 52.9%, with increasing temperature, limited by the low H content and the lack of O to the reactor. The higher heating values increased with increasing temperature from 0.77 to 1.64 MJ/Nm^3. HHV, increased by ~ 40% between 657 and 675 °C and between 675 and 698 °C, thereafter approximately 10% for each temperature rise studied. The monotonic increase resulted from improved carbon conversion at higher temperatures. The gas yield ranged from 2.26 to 2.53 Nm^3/kg and increased for the first two temperature rises then slightly decreased at 771 °C.

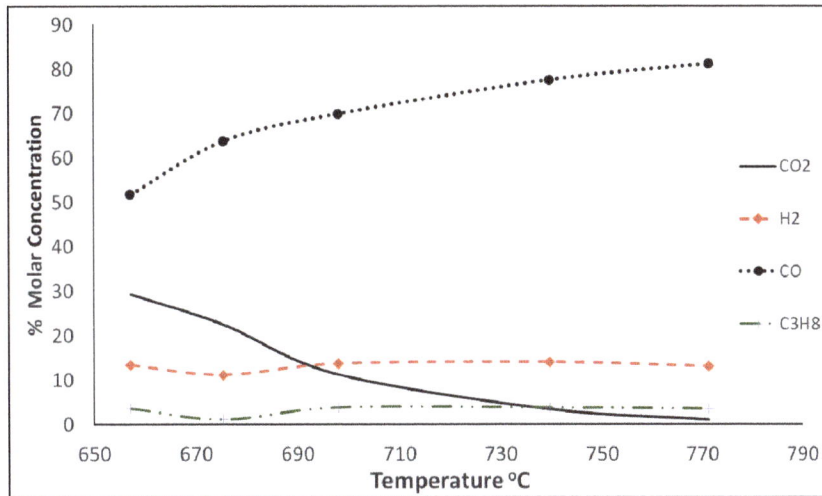

Figure 2. Effect of temperature on gas composition. Biomass feed rate: 176 g/h; ER: 0.12

Table 2. Effect of temperature on various parameters during air gasification. Biomass feed rate: 176g/h; ER: 0.13

Lower bed temperature (°C)	657	675	698	740	771
HHV (MJ/Nm^3)	0.77	1.01	1.39	1.54	1.64
Carbon conversion efficiency (%)	31.2	45.6	48.3	49.9	52.9
Gas yield (Nm^3/kg)	2.26	2.49	2.53	2.53	2.48

3.1.2 Effect of ER

The effects of ER on a number of factors were studied, with ER ranging from 0.12 to 0.25 and the reactor temperature 775 °C. While CO concentrations remained relatively constant, the concentration of H decreased, while that of CO_2 increased, with ER increasing from 0.12 to 0.25. The CO concentration was higher than that of the other gases for all ER studied. The most significant variation in the gases under study was in CO_2, with the other gases showed very little change in concentration. The carbon conversion efficiency increased from 52.9 to 89.9% as ER increased from 0.12 to 0.25, resulting from increased CO_2 (Lv et al., 2004). The calculated HHV, including N, ranged from 1.64 to 2.38 MJ/Nm^3 as shown in Table 3. The HHVs recorded when ER was varied were higher than when the effect of temperature was investigated. The gas production ranged from 2.48 to 2.73 m^3/kg. As expected, the gas production was highest for ER = 0.25.

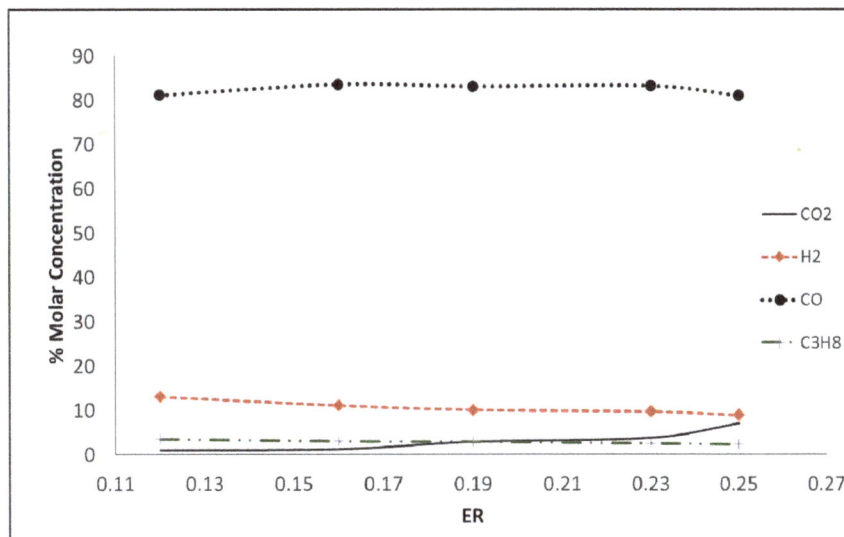

Figure 3. Effect of ER on gas composition. Biomass feed rate: 176 g/h; Temperature: 775 °C

Table 3. Effect of ER on higher heating value, carbon conversion efficiency and gas yield during air gasification Biomass feed rate: 176 g/h; Temperature: 775 °C

ER	0.12	0.16	0.19	0.23	0.25
HHV (MJ/Nm^3)	1.64	1.90	1.96	2.22	2.38
Carbon conversion efficiency (%)	52.9	63.8	68.2	79.7	89.9
Gas Yield (Nm^3/kg)	2.48	2.51	2.44	2.64	2.73

3.2 Air-Steam Gasification

3.2.1 Effect of Steam-Biomass Ratio

Analysis of the gas product was carried out at S/B ratios from 0.4 to 2.2 and a temperature of 715 ± 5 °C with ER ~ 0.12. It was difficult to maintain a fixed reactor temperature since steam gasification is endothermic process. Due to the water gas shift reaction, the concentrations of CO_2 and H_2 increased with increasing S/B ratio, while the CO concentrations decreased, as seen in Figure 4.

The carbon conversion efficiency increased when the S/B ratio increased from 0.4 to 1.3 and decreased thereafter as summarized in Table 4 summarizes the results. The highest carbon conversion efficiency of 69.7% was at S/B ratio of 1.3 with a steam flow rate of 0.216 kg/h, while the lowest value, 51.3% was at S/B = 0.4. HHV ranged from 1.95 to 2.50 MJ/Nm^3. The heating value (including N_2) reached a maximum at S/B = 1.7 due to increasing production of H_2, however the increasing CO_2 content reduces the calorific value of the product gas (Gabra, Pettersson, Backman, & Kjellstrom, 2001). The volume of gas produced ranged from 2.45 to 3.19 m^3/kg.

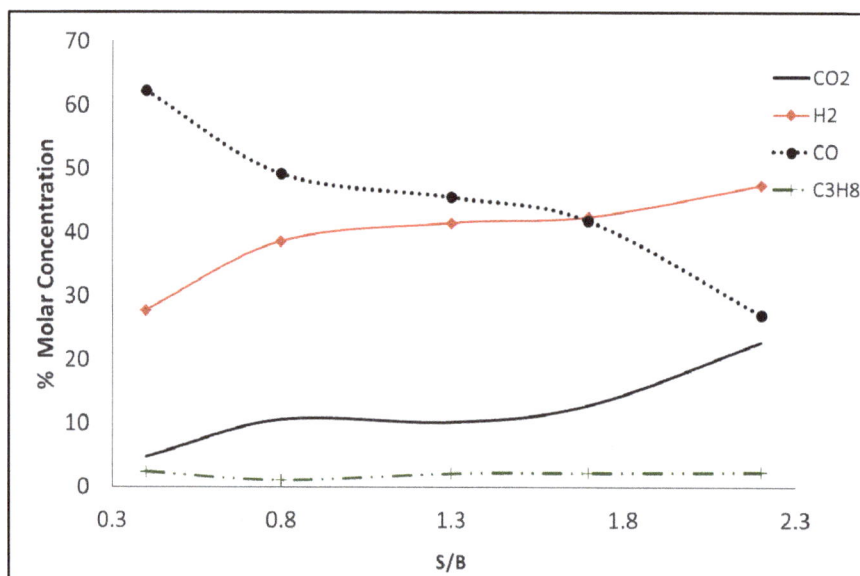

Figure 4. Effect of S/B ratio on product gas composition. Biomass feed rate: 176 g/h; Temperature: 715 °C, ER: 0.12

Table 4. Effect of S/B ratio on higher heating value, carbon conversion efficiency and gas yield during air-steam gasification for biomass feed rate: 176 g/h; ER: 0.13; Temperature: 715 °C

SB	0.4	0.8	1.3	1.7	2.2
HHV (MJ/Nm^3)	1.95	2.29	2.26	2.50	2.11
Carbon conversion efficiency (%)	51.3	63.2	69.7	66.7	54.2
Gas Yield (Nm^3/kg)	2.45	2.93	3.06	3.19	2.90

3.2.2 Effect of Temperature, ER and Steam-Biomass Ratio on H_2/CO

As shown in Figure 5, the molar H_2/CO ratio was less than 0.3 for the temperature range investigated and decreased with increasing temperature during air blown gasification. At 775 °C, all H_2/CO molar ratios were below 0.17 and decreased with increasing ER. For the air–steam blown process, as the S/B ratio increased, the H_2/CO molar ratio also increased. The H_2/CO ratios were higher than for the air-blown processes and ranged from 0.4 to 1.75. The injection of steam as gasifying agent increased the H_2/CO molar ratio because moisture promotes both steam gasification and the water gas shift reaction (Li et al., 2004; Gabra et al., 2001).

While all three runs were at different operating conditions, the HHV can be compared at similar conditions. At ER = 0.12 and 698 °C for the air blown process and at 715 °C for the steam-air gasification (S/B ratio of 0.4), the HHVs (including N_2) were 1.39 MJ/m^3 and 1.95 MJ/m^3 respectively. At this low S/B ratio, the heating value for the steam fed process was approximately 29% higher than without steam.

As discussed above, the HHV increased with increasing ER and the optimum carbon conversion efficiency was found at ER = 0.25. The increase in HHV from ER of 0.12 to 0.25 was approximately 31% for the air blown process. Since the air-steam gasification was carried out at ER = 0.12, it is likely that the HHV would increase if the ER were to be increased to 0.25, producing gas with a higher calorific value. Research on sugar cane residue (bagasse) showed that a gas generated at 3.5–4.5 MJ/Nm^3 did not present any problems when burned (Gabra et al., 2001). However, in order to obtain good burning of the gas in a turbine, combustion should be close to stoichiometric conditions. The potential of a higher heating value suggests that the product gas could be useful.

A high superficial gas velocity may cause entrainment of fine ash and carbon particles, while too low a value may result in defluidization in the reactor. While oversized particle sizes may not be fluidized and cause agglomeration (Suarez & Beaton, 2003), the feedstock under study had a low density relative to sand. No fusing of ash or ash-sand particles was observed on visually inspecting the particles after opening the reactor after each run. Hence, the possibility of reducing the fluidizing gas velocity could be considered. Gasification with N_2 as a fluidizing gas dilutes the product gas (Gil, Corella, Aznar, & Caballero, 1999). Therefore, the calorific values (not shown here) were calculated on a N-free basis to give an idea of the heating values likely to be achieved. The minimum fluidizing flow used for the experiments was approximately 15.0 L/min, achieved by adding N_2 to give the desired operating fluidization conditions. Further investigation should be carried to reduce the fluidizing flow in an effort to produce a higher calorific value producer gas.

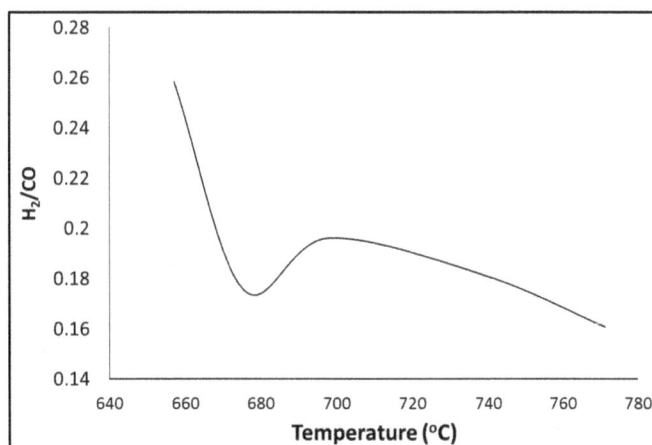

Figure 5. H_2/CO molar ratio as a function of temperature for air gasification. Biomass feed rate: 176 g/h; ER: 0.12

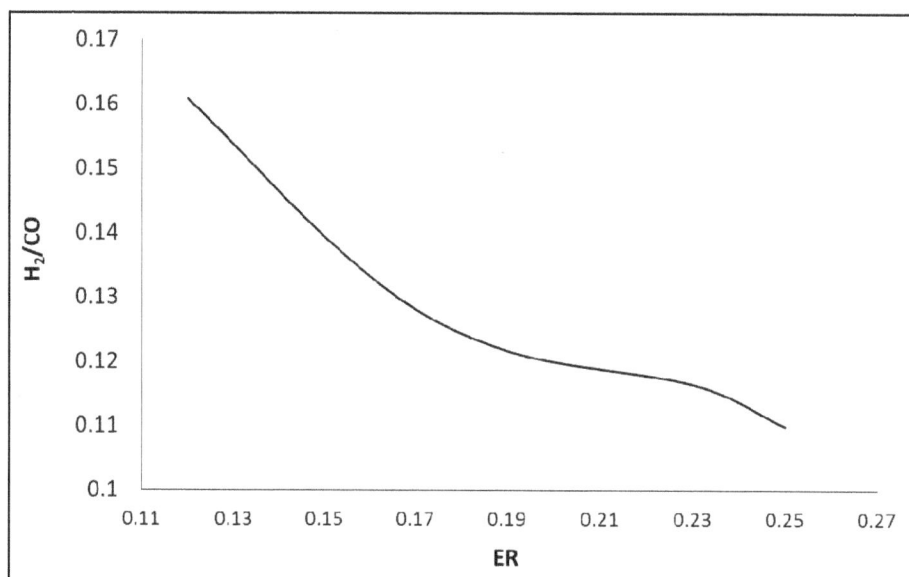

Figure 6. H$_2$/CO molar ratio as a function of ER for air gasification. Biomass feed rate: 176 g/h; Temperature: 775 °C

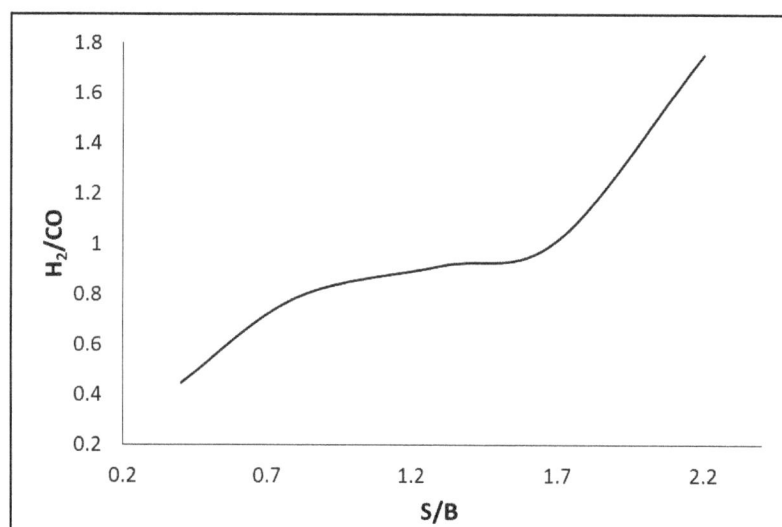

Figure 7. H$_2$/CO molar ratio as a function of S/B for air-steam gasification. Biomass feed rate: 176 g/h; Temperature: 715 °C; ER:0.12

4. Conclusion

1) High carbon ash was successfully gasified in a bubbling fluidized bed reactor at low temperatures and atmospheric pressure. Ash and woody biomass showed similar trends for gasification and product gas formation.

2) The higher heating value of the producer gas at equivalence ratios from 0.12 to 0.25 were in the range of 0.77–2.50 MJ/Nm^3 with gas yields from 2.26–3.27 Nm^3/kg.

3) The carbon conversion efficiency increased with increasing temperature, reaching a maximum at an ER of 0.25.

4) HHV increased with increasing temperature in the range 650–770 °C.

5) The heating value for the steam fed process was approximately 30% higher than without steam at otherwise similar operating conditions.

6)   The H_2/CO molar ratio increased with the addition of air-steam over the range of S/B ratio studied (0.4 to 2.2). For the air blown process the H_2/CO ratio decreased with increasing ER and with increasing temperature.

7)   No noticeable fusing of ash or ash-sand particles was observed on visual inspection after run completion.

References

Alauddin, Z. A. B. Z., Lahijani, P., Mohammadi, M., & Mohamed, A. R. (2010). Gasification of lignocellulosic biomass in fluidized beds for renewable energy development: A review. *Renewable and Sustainable Energy Reviews, 14*(9), 2852-2862. http://dx.doi.org/10.1016/j.rser.2010.07.026

Chaiprasert, P., & Vitidsant, T. (2009). Promotion of coconut shell gasification by steam reforming on nickel-dolomite. *American Journal of Applied Sciences, 6*(2), 332. http://dx.doi.org/10.3844/ajassp.2009.332.336

Cui, H., & Grace, J. R. (2007). Fluidization of biomass particles: A review of experimental multiphase flow aspects. *Chemical Engineering Science, 62*(1), 45-55. http://dx.doi.org/10.1016/j.ces.2006.08.006

Devi, L., Ptasinski, K. J., & Janssen, F. J. (2003). A review of the primary measures for tar elimination in biomass gasification processes. *Biomass and Bioenergy, 24*(2), 125-140. http://dx.doi.org/10.1016/S0961-9534(02)00102-2

Gabra, M., Pettersson, E., Backman, R., & Kjellström, B. (2001). Evaluation of cyclone gasifier performance for gasification of sugar cane residue—Part 1: gasification of bagasse. *Biomass and Bioenergy, 21*(5), 351-369. http://dx.doi.org/10.1016/S0961-9534(01)00043-5

Gil, J., Corella, J., Aznar, M. P., & Caballero, M. A. (1999). Biomass gasification in atmospheric and bubbling fluidized bed: effect of the type of gasifying agent on the product distribution. *Biomass and Bioenergy, 17*(5), 389-403. http://dx.doi.org/10.1016/S0961-9534(99)00055-0

Karamarkovic, R., & Karamarkovic, V. (2010). Energy and exergy analysis of biomass gasification at different temperatures. *Energy, 35*(2), 537-549. http://dx.doi.org/10.1016/j.energy.2009.10.022

Li, X. T., Grace, J. R., Lim, C. J., Watkinson, A. P., Chen, H. P., & Kim, J. R. (2004). Biomass gasification in a circulating fluidized bed. *Biomass and Bioenergy, 26*(2), 171-193. http://dx.doi.org/10.1016/S0961-9534(03)00084-9

Lv, P. M., Xiong, Z. H., Chang, J., Wu, C. Z., Chen, Y., & Zhu, J. X. (2004). An experimental study on biomass air–steam gasification in a fluidized bed. *Bioresource technology, 95*(1), 95-101. http://dx.doi.org/10.1016/j.biortech.2004.02.003

Mandl, C., Obernberger, I., & Biedermann, F. (2010). Modelling of an updraft fixed-bed gasifier operated with softwood pellets. *Fuel, 89*(12), 3795-3806. http://dx.doi.org/10.1016/j.fuel.2010.07.014

Natarajan, E., Nordin, A., & Rao, A. (1998). Overview of combustion and gasification of rice husk in fluidized bed reactors. *Biomass and Bioenergy, 14*(5-6), 533-546. http://dx.doi.org/10.1016/S0961-9534(97)10060-5

Pinto, F., Franco, C., Andre, R. N., Tavares, C., Dias, M., Gulyurtlu, I., & Cabrita, I. (2003). Effect of experimental conditions on co-gasification of coal, biomass and plastics wastes with air/steam mixtures in a fluidized bed system. *Fuel, 82*(15), 1967-1976. http://dx.doi.org/10.1016/S0016-2361(03)00160-1

Ryu, C., Yang, Y. B., Khor, A., Yates, N. E., Sharifi, V. N., & Swithenbank, J. (2006). Effect of fuel properties on biomass combustion: Part I. Experiments—fuel type, equivalence ratio and particle size. *Fuel, 85*(7), 1039-1046. http://dx.doi.org/10.1016/j.fuel.2005.09.019

Skoulou, V., Koufodimos, G., Samaras, Z., & Zabanioutou, A. (2008). Low temperature gasification of olive kernels in a 5-kW fluidized bed reactor for H2-rich producer gas. *International Journal of Hydrogen Energy, 33*, 6515-6524. http://dx.doi.org/10.1016/j.ijhydene.2008.07.074

Suarez, J. A. (2003). Physical properties of Cuban coffee husk for use as an energy source. *Energy sources, 25*(10), 953-959. http://dx.doi.org/10.1080/00908310390232406

Turare, C. (1997). *Biomass gasification technology and utilization.* Retrieved March 28, 2013, from http://cturare.tripod.com/ove.htm

Warnecke, R. (2000). Gasification of biomass: comparison of fixed bed and fluidized bed gasifier. *Biomass and Bioenergy, 18*(6), 489-497. http://dx.doi.org/10.1016/S0961-9534(00)00009-X

Zhang, L. B. (2011). Thermodynamic evaluation of biomass gasification with air in autotherm gasifiers. *Thermochimica Acta, 519*, 65-71. http://dx.doi.org/10.1016/j.tca.2011.03.005

11

Global Warming Potential Implications and Methodological Challenges of Road Transport Emissions in Nigeria

S. C. Nwanya[1] & I. Offili[2]

[1] Department of Mechanical Engineering, University of Nigeria, Nsukka, Nigeria

[2] Projects Development Institute (PRODA), Emene, Enugu, Nigeria

Correspondence: S. C. Nwanya, Department of Mechanical Engineering, University of Nigeria, Nsukka, Nigeria.
E-mail: stephen.nwanya@unn.edu.ng

Abstract

The purpose of this study is to examine the repercussions vehicular road transport emissions have on global warming potential (GWP), and the need to address the issue considering methodological challenges facing road transportation in Nigeria. Specific objectives of the study includes to determine the emission level in the country, to evaluate the GWP and to develop a emission mapping network on trunk A roads in Nigeria. Accurate information on these emissions is required to strengthen the mitigation and adaptation ability of the country to tackle climate change. The study relied on direct measurement technique supported by literature as well as questionnaires administered on the organised vehicle fleet operators and road traffic management agency as data gathering methods. Also, detailed analysis of questionnaires responses was carried out. Results show that road transport account for over 14% of greenhouse gases. Survey findings indicate that excessive smoke emission offence accounts for 1-2% of the annual road traffic offences in Nigeria. Using Statistical Package for Social Science (SPSS) software version 16, five fitted simple linear regression models were developed. With these fitted models it is possible to map the gas concentrations on the kilometre travelled. Examination of the National Vehicle Identification Scheme (NVIS) revealed a rise in the periodic plate number generation from yearly record of 788,169 in 2001 to 791,832 in 2009. Human capacity requirements, based on yearly Drivers Licence (DL) processed, increased by 55% between 2000 and 2010. Three mutual strategies namely renewed urban and rural road transport infrastructure availability, regular fleet maintenance and capacities building for improved behavioural change of road users were recommended to help control road transport emissions. These measures if inflexibly implemented will change the transport sector from being a major global warming risk factor to that of Eco-friendly sector.

Keywords: global warming potential, climate change, vehicle emissions, emission mapping, Nigeria

1. Introduction

In future, the rate of greenhouse gases from road transport will likely be growing faster than that from gas flaring in Nigeria. A proxy indicator of the generic cause of this emissions rate is the growing demand for personal automobile (motorcycles, tricycles, cars, trucks and tanks) ownership without a change in travel behaviours. The foregoing argument implies that vehicular emissions are inevitable corollary of the desire to meet societal mobility needs. Consequently, a large amount of greenhouse gas emissions are expected because the vehicles are operated primarily on fossil fuels. Vehicle fleets emissions are primary source of ozone precursors (Jones & Stokes, 2007). Potency of these emissions depends on quantity released and specific heat trapping property of each range (Perez et al., 2013).

The increasing intensity and heat trapping abilities of the emissions have direct consequence on warming of the earth's atmosphere referred to as global warming. Global warming potential (GWP), therefore, is a ratio of the amount of radiation a unit emission of the gas absorbs over a time frame, as compared with how much radiation a unit emission of CO_2 absorbs in the atmosphere over the same time period (Anonymous, 2012). While there is no clear consensus on global warming among scientist (Lindzen, 1992), there is a reality of its impact on ecosystem. There is also, a growing recognition that we cannot stop global warming (Igwenagu, 2011), however, we can manage their spiral effects including high global warming potential arising particularly from vehicle activities.

The purpose of this study is to examine implications of rising global warming potentials of greenhouse gas from road transport. Six physical elements of the transport sector can influence the GHG emissions: vehicle fuel efficiency, green house gas intensity of fuel used, amount of transport activity, mode of transport chosen, amount of capacity occupancy used and changes in vehicle ownership (Robinson, 1997; Briol & Guerer, 1993; Al-Naima & Hamd, 2012). For developing countries, where there is dominance of road transportation, these elements are expected to influence emission of more greenhouse gases than urbanization challenges. Under this situation, road transportation has great prospects for innovative investment opportunities in one respect. In another respect, there is air pollution as a liability. Finding sustainable solutions to the liability motivates this study.

In addition, transport sector in Nigeria is challenged by other factors including dilapidated road network, old and poorly maintained vehicle fleet and inept human capacity base. The follow-on effects of these factors can accelerate global warming. When the road is bad, it affects everybody because materials and human beings have to be moved from one location to another (Nneji, 2012). The poor state of our roads is worsened by over reliance on road transportation for mobility of goods and services resulting in avoidable accidents and traffic gridlock as shown in Figure 1a and 1b. The traffic gridlock can be compared to a queuing system. It increases vehicles fuel consumption and cost for running a normal short distance trip. For road unworthy vehicles, gridlock produces a large volume of harmful GHG emissions which contribute to global warming and can cause more short-term and localised problems, such as smog and respiratory problems (Musbau, 2012). Also, traffic congestions pose serious security risks in addition to loss of productive time. However, this study focuses on the global warming implications because of its life-threatening consequences on environment.

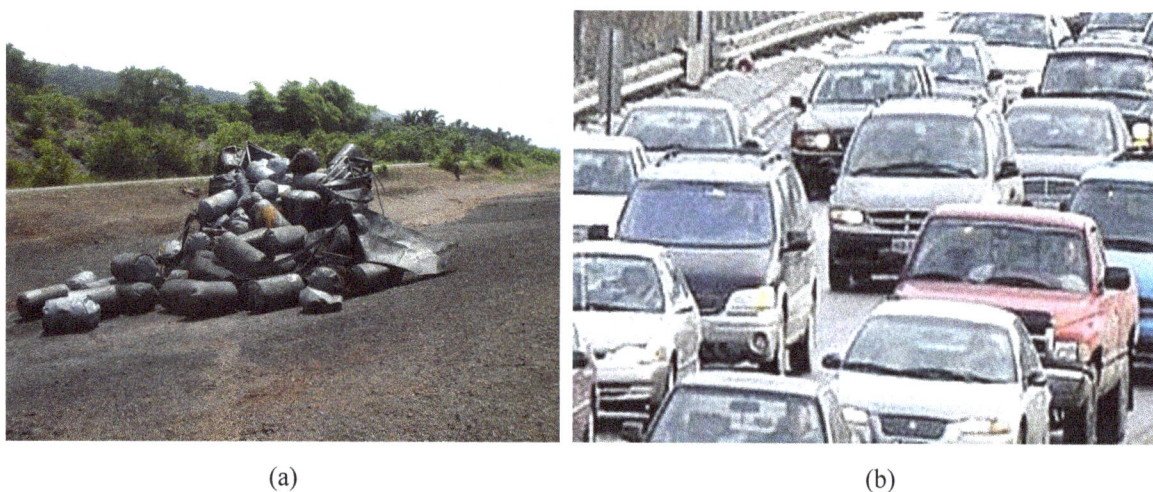

(a)   (b)

Figure 1. (a) accidental hazards due to dilapidated road; (b) traffic gridlock

The main pollutants from vehicle emissions are NO_x, CO_2, CO, SO_2, and CH_4. Pollution of ambient air with these elements all together constitutes environmental and human health hazards to the society. With increasing demand for personal automobile ownership due to impaired public transport modes, the insidious impacts of vehicle emissions are expected to continue in the future. In this context, upward trend in consumption of transport fuels and more emissions are expected (Birol & Guerer, 1993). Therefore, evaluation of the global warming potential of road transport emission is pertinent in order to evolve adaptation and mitigation strategies, especially for the associated climate change implications.

Adaptation and mitigation to climate change should be taken together because they have an inseparable link. Their linkage can be understood by the fact that adaptation practice alone cannot save a patient from continuous over dosage of a drug. There should be strategies to mitigate over dosage. In a similar sense, the fragile ecosystem is under continuous overdose of impacts of climate change and require both mitigation and adaptation strategies for continued existence. According Kane and Shogren (2000), both strategies jointly determine our risks and the costs of decreasing those risks. They each influence the timing and level required of the other (Munasinghe & Swart, 2005; Wilbanks et al., 2003) and their responses depend on social and technological constraints.

Unless inventory of vehicle emissions are evaluated and their links with social and technological constraints explored, there can be no effective control of their production. Recent studies have treated vehicle emissions both in ethnographic spread and content analysis. In the context of ethnographic occurrence, (Abam & Unachukwu, 2009; Utang & Peterside, 2011; Ndoke et al., 2012; Moen, 2012; Musbau, 2012) focused on Calabar, Port Harcourt, North Western Nigeria, Abuja and Lagos cities, respectively. These studies employed sampling technique of USEPA standard using portal analyser to estimate road transport pollutant indicators at spots or sampling points that designate high congestion and dense population. In terms of report content, Ojolo et al. (2007) presented a generalised analysis of the impact of vehicle emissions on human health in Nigeria and Utang Peterside (2011) identified spatio-temporal variations in vehicle emissions between traffic peak and off peak periods in Port Harcourt city. These studies have their strength and weaknesses as well as criteria for selection of sampling locations that are inimical to emission from mobile sources.

It is common knowledge that vehicle emission concentrations vary with engine operating mode, which include idling, decelerating, cruising and accelerating. The present literature on the subject so far has treated and evaluated emissions from automotive at traffic jam and sampling points alone. There seems no documentation exists that covers major road networks across all geographic areas. In order to close this gap, the emission mapping of traffic pollutants along the roads between Abuja and States capital is important and addressed in this study.

The objectives of this study include determining the emission level in the country, evaluating the GWP and developing an emission mapping network on trunk "A" roads in Nigeria. In carrying the study, relevant data were obtained through questionnaires and interviews methods. Contacts were made to competent personnel of the Federal Road Safety Commission (FRSC) (a typical traffic management system) and the organised vehicle fleet operators while the obtained data were used to compute GWPs indices. Unlike the treatment in the present literature on the subject GHGs emission this study developed a mapping methodology. This methodology is a consistent approach for calculating the GHGs emission along the major trunk A roads connecting Abuja city with State capitals. The mapping is with a view to identifying road networks that lead to higher emissions and capacity expansion options for the sector. The approach is robust and exploits travel behaviour (cruise, accelerating and decelerating) applicable for intercity connections.

1.1 Background Information and Justification for the Work

Globally, transportation service is conducted mainly by water ways, road, railways and airways. These modes collectively provide safer and faster ways for movement of goods and services. But, transportation in Nigeria is a harrowing experience due to over reliance on road transport vehicles. Road transport is the most popular; accounting for 75% of mobility needs and has widest spread of transport network in Nigeria. The country has 195,000 km national road network, with federal roads accounting for 17% (Nigerian Engineer, 2012), current vehicle population exceeding seven million (Agbo, 2011) and average density of 11 vehicles per kilometre (Musbau, 2012).

The population of Nigeria is 140 million people from 250 ethnic nationalities. The ethnic nationalities have been grouped into 36 States as shown in Figure 2 with Abuja as federal capital for administrative convenience and political governance. The roads linking the States with Abuja have high traffic density, which informed their choice for this study. Abuja is Nigeria's fastest growing city with a per annum growth rate of 5% and low industrial development (Moen, 2012). The overall rate of urbanization in Nigeria is 5.6%. Therefore, high level socio-economic and political interactions between the States and Abuja necessitate intensive transportation services. However, lack of railways and waterways as well as unreliable airways systems connecting the States and federal capital have resulted in heavy vehicle activity with the associated GHG emissions.

The actual and potential effects of concentrating on single mode of transport, according to Robinson (1997), are large-global climate change, ozone depletion, smog, congestion, noise and urban sprawl. Estimation of the level of concentration of gas emissions and associated impacts like global warming under heavy vehicular activity is the main thrust of this study.

The vehicle activity is defined here as the distance covered by vehicles between the States and federal capital routes. The associated emissions have global warming potential implications. Global warming potential implies increase irradiative capacity or ability to trap ambient heat. These implications in addition to deplorable state of the roads make road travel unpleasant and stressful or sometimes result in human health problems. The argument against this worrisome experience has led to short term recommendations for strict control measures such as restriction on importation of old vehicles and elimination of lead concentration in gasoline. But, these measures hardly offered reasonable relief and for this reason complementary solutions are required. Long term measure will include emission mapping that will ensure optimization of the potency level of the GHGs.

Figure 2. Map of Nigeria showing the 36 States and Abuja capital

2. Challenges of the Road Transportation System in Nigeria

A comparison of modes of powering road transport vehicles clearly indicates that whereas electric mode ranks second to internal combustion engines in terms of cost-benefits, the former remains the best in terms of air pollution effects. Electric vehicle offers an improved overall thermodynamic efficiency (Muneer et al., 2011). However, electric vehicles are uncommon in Nigeria and replacement of internal combustion engine with electric type involves huge capital investment in infrastructure and capacity building. Also, the growth potential of road transport has been associated with corresponding increased energy intensity. Energy intensity is defined here as the energy use per vehicle kilometre in the case of cars, energy use per tonne kilometre for goods vehicle, rail, marine and air freight, and energy per passenger kilometre for bus, air and rail transport (Michaelis, 1997).

Efficient and safe road transportation is among the missing critical success factors in the drive towards a productive economy in Nigeria. This problem undermines sustainable development, particularly economic transformation that stems from mobility of human, goods and services. Ineffective transport system in Nigeria has been attributed to low quality of human capacity in the sector, increasing number of old and poorly maintained vehicle fleet, over stretching of road and its substitution for rail and water ways. How can these issues be handled with a view to innovatively tap the job creative opportunities of road transport sector?

2.1 Capacity Building and Training for Promotion of Reliable Road Transport System

The road transportation management in Nigeria is challenged by unskilled human capacity base. The effect of this problem on the society includes high accident rate and 70% of it is attributed to ineptitude. Without qualified and tested human capacity base, there can be no effective road transportation system. Capacity building is prescribed here as a part of the way forward.

Capacity building in the context of road transportation sector involves a process of assisting people to develop their technical and decision making skills for getting it right when faced with road challenges. The road transport sector like most other sectors should benefit from the abundant human resources in the country. The jobs available in the sector are such that the seeker creates them and can just do that only if the requisite skills have been embedded. In view of this reason, the human capacity aspect of the road transport administration can, as a general rule of management, only increase productivity if the capacity is enhanced. Are there possibilities for this purpose in contemporary Nigeria?

2.2 Fleet Maintenance, Elimination of Aging and Air Polluting Vehicles

The social and economic consequences of heavy road vehicular traffic has remained an under- surface problem in Nigeria. Nigeria is highly dependent on second-hand used imported vehicles that place great burden on the environment in terms of energy consumption and air pollution. Imported brand new vehicles that have high fuel efficiency are expensive and mostly affordable by government agencies and few corporate organizations. The above stated trend of vehicle importation has remained difficult to change since the few assemble plants closed shops in Nigeria. Introduction of road user charges such as excessive emission charges, fuel taxes and vehicle import taxes rated in proportion to age of manufacture will address the intractable emission level. The argument for these charges is to let the polluter pay cost for damages caused. For effective implementation of emission charge, it should be pressed on vehicles that fail a yearly tailpipe test.

Apart from the age of vehicles, improper fuel combustion techniques precipitated by untrained operators/ drivers also accounts for the rising social and economic costs of fleet management. Muneer et al. (2011), has reported that 23% of energy losses in vehicles are due to frequent standby and breaking by drivers. This leads to more CO_2 emissions.

Another alternative way forward includes adoption of railway system as widely accepted means of mass freight and passenger transport, use of electric vehicles and renewable energy-based vehicles that will facilitate green technology transformation in Nigeria.

2.3 Urban and Rural Transport Infrastructure Availability

Road construction and maintenance in both urban cities and rural communities are twin indicators of good governance. But, they are lacking in Nigeria, while dilapidated tarred and eroded laterite-paved roads embellish the traffic ways.

The problem can be solved by involving private partnership with government in road construction and maintenance. Urban and rural road construction will reduce traffic congestion in the light of increasing vehicle ownership in the economy. Therefore, a strategy will be adopted through which the private sector recovers the investment in road construction and maintenance such as through road toll concession.

Three mutual strategies that need to be reinforced to ensure control of road transport emission and at the same time reduce the challenges are: renewed urban and rural transport infrastructure availability, regular fleet maintenance, consistent check on aging and air polluting vehicles, and capacity building for improved behavioural change of road users. These measures if inflexibly implemented will change the transport sector from being a major global warming risk factor to that of Eco-friendly sector.

3. Materials and Methods

Quantifying the current emissions from road transport is a crucial first step to determining the environmental impact of the sector. The approach used followed estimation of emission quantities and questionnaire administration. As for the basic emissions quantification, we have followed aforementioned instantaneous measurements conducted in Nigeria, but have modified these wherever this seemed appropriate so as to avoid built-in biases. The instantaneous measurements monitored directly air quality parameters through a Testo 350XL Emission Analyzer. The monitoring was conducted for a minimum of 72 hours at road intersections having records of perennial traffic congestions. Emissions concentration data were collected at hourly intervals for peak and off-peak periods. The concentrations are measured in parts per million. The monitoring process followed the pattern of exhaust testing.

In this regard, the average of the common emission values is calculated with at least one more figure than that of the data as follows:

$$\bar{y} = \frac{1}{n}\sum y_i \qquad (1)$$

Where n = total number of sample observations, i = 1 to 3 parameters,

To calculate the greenhouse gases emissions from transport sector, the following are necessary:

- Obtain emission concentration factor of each of GHG pollutants and the transport activity data for the national territory

- The activity data is obtained from measurements, while emission factors are instantaneous values in concentration form, parts per million (ppm)

- The instantaneous values are converted from ppm to an equivalent mass values in kilogram

- Using the activity data and emission factors, then calculate emission concentrations generated that are specific to each GHG

- The GHG emissions are reported in terms of CO_2 equivalent. The CO_2 equivalent is determined by multiplying the amount of emissions of a particular gas by its specific global warming potential index. Table 1 shows specific global warming potential indices of some of the gases.

- Also, an assumption is made that the concentration of pollutants in atmosphere is influenced by weather conditions which are different in space and time (Utang et al., 2011).

To reduce all values to single unit, the concept million metric tons of carbon equivalent (mmtce) is used (GEF, 2010).

Table 1. Global warming potential indices of some of greenhouse gases

Gas	Lifetime	Global warming potential time horizon		
	Years	20 years	100 years	500 years
Methane CH_4	12	72	25	7.6
Nitrous Oxide N_2O	114	289	298	153
Hydrofluorocarbon HFC-23	270	1200	14,800	12,200
Hydrofluorocarbon HFC-134a	14	3800	1430	435
Sulphur Hexafluoride	3200	16300	22,800	32,600

Source: GEF, 2010.

4. Results and Discussion

For the purpose of this study, 36 heavily trafficked routes that link the 36 States of Nigeria with Abuja were used to represent the traffic activity. A questionnaire survey considered predominant automotive vehicle types in Nigeria irrespective of model and fuel types. The investigation showed that cars, buses and jeeps are common on our roads. Also, results of the questionnaire survey show that some of the respondents spend 30 to 60 minutes of their travel time at traffic congestion. Human capacity requirements for transport sector, based on yearly Drivers Licence (DL) processed, increased by 55% between 2000 and 2010. This implies a growth rate of 5.5% if spread evenly within the range of period being considered. This growth rate is significant in view of rising number of personal automobile ownership. It can be inferred that with the 5.5% and more vehicles operated on fossil fuel, there will be a corresponding increase in amount of emissions resulting to high GWPs. The survey analysis also, showed that overall annual smoke emission offences accounts for 1% to 2% of road traffic offence.

For the objective measurement of exhaust gas emissions from these vehicles, the study adopted direct measurements monitored by Abam and Unachukwu (2009), Moen (2012), Utang and Peterside (2011). The decision to adopt these values was taken because the authors followed international standard practices similar to the strategy proposed for this study. Also, the following: NO_x, CO_2, CO, SO_2, and CH_4 were monitored at different geographical zones with peak and off-peak periods within same range of ambient conditions which fairly represent national conditions. Table 2 shows some of the different readings.

The concentration levels shown in Table 2 collaborates with the survey questionnaire findings that excessive smoke emission offence accounts for 1-2% of the annual road traffic offences in Nigeria. Also, road transport account for over 14% of national greenhouse gas emissions.

The aforementioned values are constrained to increase when examined along side the rising rate of vehicle ownership, both government and private in the country. Table 3 shows the ownership structure through periodic National Vehicle Identification Scheme (NVIS) registration.

Table 2. Some of the hourly ambient gas concentrations of vehicular emissions in Nigeria

City	CO (ppm)	No. observations	SO$_2$ (ppm)	No. observations	NO$_2$ (ppm)	No. observations
Abuja	24	20 morning	0.6	22 morning	N/A	N/A
Abuja	16	23 afternoon	0.3	24 afternoon	N/A	N/A
Calabar Day 1	6.177	9 morning	0.062	9 morning	0.05	9 morning
Calabar Day 2	5.7	9 morning	0.07	9 morning	0.05	9 morning
Calabar Day 3	5.8	9 morning	0.07	9 morning	0.05	9 morning
Port Harcourt	5.5	4 peak morning	0.004	4 peak	0.43	4 peak
Port Harcourt	16.07	4 peak evening	N/A	N/A	0.95	4 peak

Source: Abam and Unachukwu (2009), Moen (2012), Utang and Peterside (2011).

N/A = not available.

Table 3. National vehicle identification scheme (NVIS) showing vehicle ownership structure in Nigeria

Organization	Year-2001	2002	2003	2004	2005	2006	2007	2008	2009
Military/Param.(MV)	226	0	10,365	839	2,366	429	2,360	2,449	2,205
Military/ Param.(MC)	0	0	344	30	12	319	1,060	202	354
Diplomatic (MV)	3,427	0	682	413	1,084	437	269	166	90
Diplomatic (MC)	5	0	0	0	0	0	1	1	5
Federal Govt (MV)	13,532	26,096	12,721	2,576	2,515	3,203	4,311	3,556	3,354
Federal Govt. (MC)	176	266	1,648	199	744	4,440	770	1,692	487
State Govt. (MV)	317	863	48,975	3,911	5,068	5,734	7,399	8,007	8,456
State Govt. (MC)	0	0	7,985	623	1,300	1,545	1,502	1,202	1,384
Local Govt. (MV)	0	124	14,136	811	744	788	480	2,508	1,299
Local Govt. (MC)	0	0	3,251	142	285	113	166	653	204
Private (MV)	400,014	383,750	238,913	206,102	167,032	178,061	204,887	231,756	252,126
Private (MC)	163,822	187,489	248,216	203,908	217,862	308,228	284,206	351,247	343,888
Commercial (MV)	124,498	91,631	59,202	49,558	57,526	53,322	51,901	71,064	90,937
Commercial (MC)	82,152	109,608	94,345	92,002	72,541	97,133	66,792	87,499	87,043
Total	788,169	799,827	740,783	561,114	529,079	653,752	626,104	762,002	791,832

Source: Federal Road Safety Corps (FRSC), Wuse Zone 2, Abuja, Nigeria.

The prior stated gaseous concentrations in Table 2 are emitted as exhaust gases from the automotive engine and into the atmosphere through the exhaust ports, exhaust manifold, exhaust pipe, muffler, and tail pipe. They are still average point data (as calculated using Equation 1) and monitored within a kilometre range. To calculate the instantaneous gases emitted in mass units, a measured exhaust flue gas volume flow rate ($V_{exh.}$) of 1200 CFM (0.564 m^3/s) is used. To obtain the mass units for each of NO$_2$, CO and SO$_2$, these steps were followed:

$$V_{exh.} = 3600 \times 0.564 [m^3/h] \tag{2}$$

$$M_{poll.i} = Con_{poll.i} \times 10^{\wedge}-6 \times \rho_{pool.i} \times V_{exh.} [kg/h] \tag{3}$$

Where:

$M_{poll.i}$ = mass emission of pollutant, I = 1 to n, Con_{poll} = concentration pollutant in parts per million, P = density of pollutant, g/dm^3, $V_{exh.}$ = exhaust volume flow rate m^3/s, m = metre, h = hour and 3600 multiplies 0.564 m^3/s to get the cubic metre per hour.

Each quantity of gas emitted can be appraised with traffic activity or vehicle kilometre travelled (VKT) and vehicle type as well as vehicle population. The impacts of the three independent variables on emission concentration and GWP are investigated among the 36 States in this study as shown in Table 4.

Table 4. Effect of traffic activity and vehicle population on gaseous concentration

City	VKT	VP	a	b	c	d	e	f
Abeokuta	740	116895	16616.7	1.14256	0.000772	1.94E+09	133559.6	90.2219
Akure	700	86273	15718.5	1.0808	0.00073	1.36E+09	93243.86	62.98792
Asaba	404	137615	9071.82	0.623776	0.000421	1.25E+09	85840.93	57.98711
Awka	440	134227	9880.2	0.67936	0.000459	1.33E+09	91188.45	61.59945
Bauchi	445	63995	9992.475	0.68708	0.000464	6.39E+08	43969.68	29.70232
Benin city	450	99540	10104.75	0.6948	0.000469	1.01E+09	69160.39	46.7191
Birin Kebbi	573	31681	12866.72	0.884712	0.000598	4.08E+08	28028.56	18.9338
Calabar	857	108551	19243.94	1.323208	0.000894	2.09E+09	143635.6	97.02842
Damaturu	757	52182	16998.44	1.168808	0.00079	8.87E+08	60990.74	41.20035
Dutse	512	35109	11496.96	0.790528	0.000534	4.04E+08	27754.65	18.74877
Bayelsa	930	148152	20883.15	1.43592	0.00097	3.09E+09	212734.4	143.706
Ebonyi	695	148986	15606.23	1.07308	0.000725	2.33E+09	159873.9	107.9977
Ado-Ekiti	600	85062	13473	0.9264	0.000626	1.15E+09	78801.44	53.2318
Enugu	595	105370	13360.73	0.91868	0.000621	1.41E+09	96801.31	65.39104
Ibadan	659	182201	14797.85	1.017496	0.000687	2.7E+09	185388.8	125.2335
Ikeja	879	463349	19737.95	1.357176	0.000917	9.15E+09	628846.1	424.797
Ilorin	500	56135	11227.5	0.772	0.000522	6.3E+08	43336.22	29.2744
Jalingo	591	31688	13270.91	0.912504	0.000616	4.21E+08	28915.43	19.5329
Gombe	501	53829	11249.96	0.773544	0.000523	6.06E+08	41639.1	28.12797
Jos	313	92762	7028.415	0.483272	0.000326	6.52E+08	44829.28	30.28299
Kaduna	180	121926	4041.9	0.27792	0.000188	4.93E+08	33885.67	22.89039
kano	442	199154	9925.11	0.682448	0.000461	1.98E+09	135912.2	91.81119
Katsina	533	44307	11968.52	0.822952	0.000556	5.3E+08	36462.53	24.6311
Keffi	80	151602	1796.4	0.12352	8.34E-05	2.72E+08	18725.88	12.64967
Gauzu	660	87627	14820.3	1.01904	0.000688	1.3E+09	89295.42	60.32067
Lokoja	138	76142	3098.79	0.213072	0.000144	2.36E+08	16223.73	10.95942
Madugiri	908	88739	20389.14	1.401952	0.000947	1.81E+09	124407.8	84.03974
Makurdi	323	44591	7252.965	0.498712	0.000337	3.23E+08	22238.07	15.02222
Minna	117	77640	2627.235	0.180648	0.000122	2.04E+08	14025.51	9.474487
Oshogbo	428	96893	9610.74	0.660832	0.000446	9.31E+08	64029.99	43.25342
Owerri	733	127442	16459.52	1.131752	0.000765	2.1E+09	144232.7	97.43183
P.Harcourt	830	152396	18637.65	1.28152	0.000866	2.84E+09	195298.5	131.9277
Sokoto	703	91103	15785.87	1.085432	0.000733	1.44E+09	98886.11	66.79936
Umuahia	700	115638	15718.5	1.0808	0.00073	1.82E+09	124981.6	84.4273
Uyo	828	77824	18592.74	1.278432	0.000864	1.45E+09	99492.69	67.20912
Yola	855	55281	19199.03	1.32012	0.000892	1.06E+09	72977.55	49.29766

Source: authors.

Where: VKT = vehicle kilometre travelled, VP = vehicle population, a = VKT* instantaneous ppm of CO, b = VKT * instantaneous ppm of SO_2, C = VKT * instantaneous ppm of NO_2, d = VP * instant. Ppm of CO, e = VP * ppm of SO_2, and f = VP * ppm of NO_2.

Using Statistical Package for Social Science (SPSS) software version 16, a five fitted simple linear regression models were developed. With these fitted models it is possible to map the gas concentrations on the kilometre travelled. Also, the fitted models show that gas concentrations are functions of population of automotive vehicles. Equations 4 and 5 demonstrate relationship between VKT and concentration, while equations 6, 7 and 8 are for VP and concentration.

$$C\hat{O} = 22.455 V\hat{KT} \quad ; \quad r^2 = 1 \tag{4}$$

$$S\hat{O}_2 = 0.001 V\hat{KT} \quad ; \quad r^2 = .68 \tag{5}$$

$$C\hat{O} = -5.258 \times 10^{\wedge}8 + 1.7854 \times 10^{\wedge}4 V\hat{P}; \quad r^2 = .75 \tag{6}$$

$$S\hat{O}_2 = -3.8328 \times 10^{\wedge}4 + 1.26 V\hat{P} \quad ; \quad r^2 = .796 \tag{7}$$

$$N\hat{O}_2 = 25.988 + 0.001 V\hat{P}; \quad r^2 = 0.829 \tag{8}$$

The developed models are useful for forecasting the gas concentrations in once the VKT and VP are known. With known concentration, emission factors can be estimated. Using the emission factor and activity data the GWP specific to each GHG is computed.

The GWP of each GHG is defined in relation to a given weight of CO_2 and for a time period. Currently, GWP of gases are expressed in CO_2 equivalents and it makes comparison of the potential effects of other gases possible (GEF, 2010). In this work, tons CO_2 eq /km for three commonly available automobiles: cars, buses and jeeps in Nigeria were calculated based on their sample population. The result is shown in Figure 3. From Figure 3, it can be proved that increasing vehicular emissions have GWP and climate change implications on human health and the environment. Meanwhile in this study, reduced severity of impacts can be realized through improved changes in vehicle characteristics (efficiency), road characteristics (increase occupancy per peak hours) and travel preference mode.

5. Conclusion

In this study, a global warming potential implication of vehicle emissions and methodological challenges faced in a road dominated mode of transport has been examined. The greenhouse emissions have pollution effects and are the main cause of global warming and climate change. In carrying the study, relevant data were obtained through questionnaires and interviews methods.

Results show that road transport account for over 14% of greenhouse gases. The questionnaire analysis indicates that excessive smoke emission offence accounts for 1-2% of the annual road traffic offences in Nigeria. Under the National Vehicle Identification Scheme (NVIS) the periodic plate number generation rose from yearly record of 788,169 in 2001 to 791,832 in 2009. More so, the human capacity requirements, based on yearly Drivers Licence (DL) processed, increased by 55% from 2000 to 2010. Using Statistical Package for Social Science (SPSS) computer software, a five fitted simple linear regression models were developed. With these fitted models it is possible to map the gas concentrations on the kilometre travelled. Road transport emissions have great climate change implications on human health and the environment. In Nigeria, challenges to efficient transportation have been attributed to low quality of human capacity in the sector, increasing number of old and poorly maintained vehicle fleet, over stretching of road and its substitution for rail and water ways.

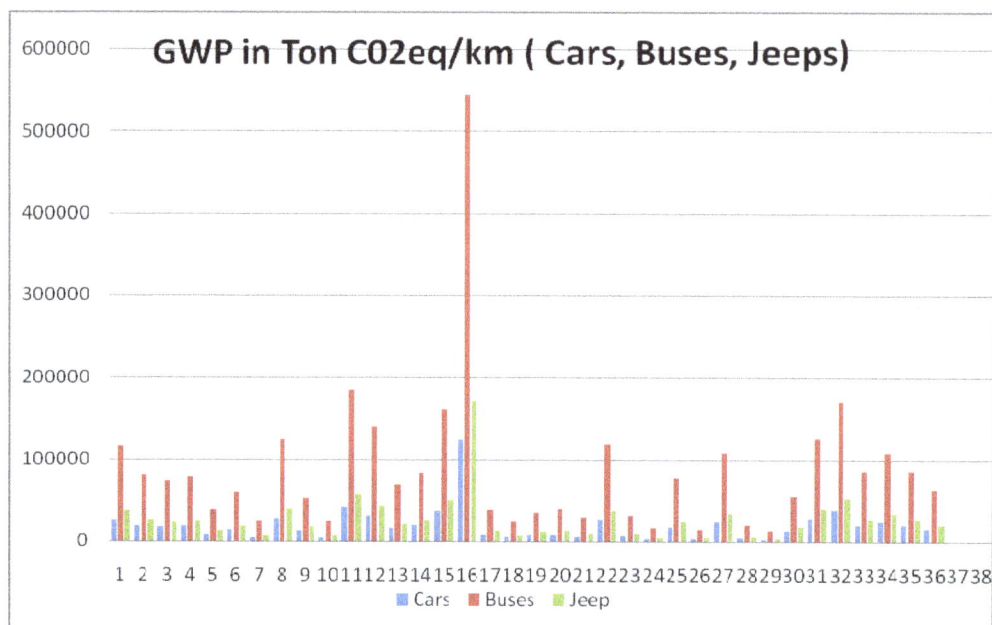

Figure 3. GWP in Ton CO_2 eq/km for cars, buses and jeeps

Three mutual strategies are needed to control road transport emission and at the same time reduce the challenges are: renewed urban and rural transport infrastructure availability, regular fleet maintenance and capacity building for improved behavioural change of road users. Also, this study recommends a yearly tailpipe emission testing on all vehicles which qualifies them to ply on public roads.These measures if inflexibly implemented will change the transport sector from being a major global warming risk factor to that of Eco-friendly sector.

Information obtained from this study will help policy makers to formulate road pricing systems and plan for green technological transformation.

References

Abam, F. I., & Unachukwu, G. O. (2009), Vehicular emissions and air quality standards in Nigeria. *European Journal of scientific research, 34*(4), 550-560.

Agbo, C. O. A. (2011). Recycle materials potential of imported used vehicles in Nigeria. *Nigerian Journal of Technology, 30*(3), 118-129.

Al-Naima, F. M., & Hamd, H. A. (2012). Vehicle traffic congestion estimation based on radio frequency identification and wireless sensor networks. *International J. Engineering Business Management, 4*(30), 1-8.

Anonymous. (2012). *Global warming potentials*. Retrieved February 20, 2012, from www.eeocw.org/get-involved/global-warming-potential

Birol, F., & Guerer, N. (1993), Modelling the transport sector fuel demand for developing economies. *Energy Policy, Butterworth-Heinemann limited, United Kingdom*, 1163-1172.

Global Environment Facility (GEF). (2010). *Manual for calculating greenhouse gas benefits for global environment facility transportation projects* (pp. 1-31). Washington D.C.: GEF/C.39/Inf.16.

Igwenagu, C. M. (2011). Principal component analysis of global warming with respect to CO_2 emission in Nigeria: an exploratory study. *Asian Journal of Mathematics and Statistics, 4*(2), 71-80. http://dx.doi.org/10.3923/ajms.2011.71.80

Jones & Stokes. (2007). *Addressing Global Warming (Climate Change) in CEQA and NEPA Documents in the Post AB 32 Regulatory Environments*. Retrieved July 27, 2012, from www.climatechangefocusgroup.com

Kane, S., & Shogren, J. (2000). Linking adaptation and mitigation in climate change policy. *Climate change, 45*, 75-102. http://dx.doi.org/10.1023/A:1005688900676

Lindzen, S. R. (1992). *Global warming: origin and nature*. Retrieved March 7, 2013, from www.cato.org/pubs/regulation/reg15n2g.html

Michaelis, L. (1997). Transport sector-strategies markets, technology and innovation. *Energy policy, 25*(14-15), 1163-1171. http://dx.doi.org/10.1016/S0301-4215(97)00108-0

Moen, E. (2012). Vehicle emissions and health impacts in Abuja, Nigeria.

Munasinghe, M., & Swart, R. (2005). *Primer on climate change and sustainable development: facts, policy analysis applications.* Cambridge: Cambridge University Press. http://dx.doi.org/10.1017/CBO9780511622984

Muneer, T., Celik, A. N., & Canskan, N. (2011), Sustainable transport for a medium sized town in Turkey- a case study. *Sustainable Cities and Society, 1*, 23-37. http://dx.doi.org/10.1016/j.scs.2010.08.004

Musbau, R. (2012). Traffic gridlock, road safety and traffic radio. Business day Newspaper, July 12, 14.

Ndoke, P. N., Akpan, U. G., & Kato, M. E. (2012), Contributions of Vehicular Traffic to Carbon Dioxide Emissions in Kaduna and Abuja, Northern Nigeria. *Leonardo Electronic Journal of Practices and Technologies*. Retrieved May 23, 2012, from http://webmail.academicdirect.org/cgi-bin/openwebmail

Nneji, F. (2006). The benefits of good transport system in Nigeria. Business day Newspaper.

Perez, P., Ortega, M., Martin, B., Otero, I., & Monzon, A. (2013). Transport planning and global warming. Retrieved March 15, 2013, from www.cdn.intechweb.org/pdfs/12174.pdf

Robinson, R. (1997). Transportation demand Management in Canada: an overview. *Energy Policy, 25*, 1189-1191. http://dx.doi.org/10.1016/S0301-4215(97)00120-1

The Nigerian Engineer. (2012). 2011 in reminiscence: The Nigerian Society of Engineers' 2011 conference and annual general meeting. *Magazine of the Nigerian Society of Engineers*, 4-6.

Utang, P. B., & Peterside, K. S. (2011). *Spatio-temporal variations in urban vehicular emission in Port Harcourt city, Nigeria.* Retrieved March 7, 2012, from http://dx.doi.org/10.4314/ejesm.v4i2.5

Wilbank, T. J., Kane, S. M., Leiby, P. N., Perlack, R. D., Settle, C., Shogren, J. F., & Smith, J. B. (2003). Integrating mitigation and adaptation – possible responses to global climate change. *Environment, 45*, 28-38. http://dx.doi.org/10.1080/00139150309604547

12

Quantifying Space Heating Stove Emissions Related to Different Use Patterns in Mongolia

Randy L. Maddalena[1], Melissa M. Lunden[1], Daniel L. Wilson[1,2], Cristina Ceballos[1,2], Thomas W. Kirchstetter[1], Jonathan L. Slack[1] & Larry L. Dale[1]

[1] Lawrence Berkeley National Laboratory, Berkeley, CA, USA

[2] University of California, Berkeley, CA, USA

Correspondence: Larry L. Dale, Lawrence Berkeley National Laboratory, Sustainable Energy Systems Group, Energy Efficiency Standards Group, Lawrence Berkeley National Laboratory, 1 Cyclotron Road MS 90R4000, Berkeley CA 94720, USA. E-mail: LLDale@lbl.gov

Abstract

A major source of particulate matter pollution in Mongolia's capital, Ulaanbaatar, is emissions from traditional coal-burning space-heating stoves. Significant investment has been made to replace traditional highly polluting heating stoves with improved low-emission high-efficiency stoves. Performance testing that has been undertaken to support the selection of replacement stoves is typically based on manufacturers' recommended operating procedures, which may not be representative of the operating procedures used in homes. The objective of this research is to evaluate factors that influence stove emissions under typical field operating conditions. A highly-instrumented stove testing facility was constructed to allow for rapid and precise adjustment of factors influencing stove performance. Tests were performed using one of the improved stove models currently available in Ulaanbaatar. Complete burn cycles were conducted with coal from the Ulaanbaatar region using various startup parameters, refueling conditions, and fuel characteristics. Measurements were collected simultaneously from undiluted chimney gas, diluted chimney gas, and plume gas drawn from a dilution tunnel above the chimney. Ignition events lead to increased PM emissions with more than 98% of PM mass emitted during the startup and refueling process. However, emissions during refueling are of particular interest, both because refueling is common and because refueling associated emissions appear to be very high. CO emissions are distributed more evenly over the burn cycle, peaking during ignition and late in the burn cycle. We anticipate these results being useful, in combination with behavioral surveys, for quantifying public health outcomes related to the distribution of improved stoves and to identify opportunities for improving and sustaining performance of the new stoves.

Keywords: air pollution, coal, laboratory testing, particulate matter, Ulaanbaatar

1. Introduction

Air pollution levels in Ulaanbaatar, Mongolia's capital, are among the highest in the world (World Bank, 2011). The primary source of particulate matter pollution in and around Ulaanbaatar is wind-blown dust and combustion products related to transportation, energy, and in-home heating and cooking (Davy et.al., 2011; Lodoyasamba & Pemberton-Pigott, 2011). The traditional coal-fired space heating stoves used in the Ger (tent) neighborhoods around Ulaanbaatar are a major source of particulate matter pollution during the winter months (Allen, et al, 2013; Iyer, Wallman & Gadgil, 2010).

Significant investment has been made to replace traditional space heating stoves with improved low-emission high-efficiency stoves. Selection of these high-performance heating stoves is based on scripted performance and emission testing protocols that are often based on manufacturers' recommended operating procedures. These idealized test conditions demonstrate the stoves' optimal performance but they do not account for non-ideal stove operation by users, and therefore manufacturers' results may not be representative of true in-field performance and emissions.

Pemberton-Pigott (2011) summarized a large number of stove performance tests conducted at the Stove Emissions and Efficiency Testing (SEET) Laboratory in Ulaanbaatar, noting that stove performance was impacted by how the stoves were actually being used compared to the recommended operating method. Furthermore, Lobscheid, Fitts, Lodoysamba, Maddalena & Dale (2014) report that field operation of improved stoves varies significantly from manufacturers' recommended operating methods, but emission measurements for the range of operating methods observed in the field are not available.

Stove performance in the field can vary widely due to differences in environmental conditions and operating behaviors. For example, a migrating pyrolytic front stove, often called a "TLUD" (Top-Lit/Up-Draft) is designed to have a stack of fuel batch-loaded into a combustion chamber and then ignited from the top (Figure 1). Improved emissions performance for this style of stove is achieved by the slow downward migration of the pyrolytic front. However, this design requires the stove to cool below the fuel ignition temperature before refueling. If cold fuel is added on top of hot embers, the fuel stack could ignite from below resulting in an updraft scenario marked by reduced performance and a significant increase in emissions.

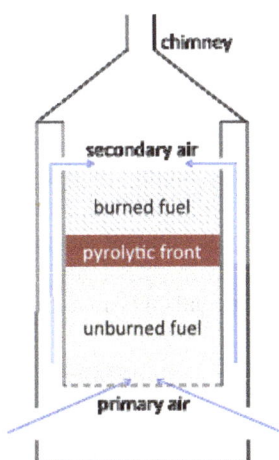

Figure 1. Functional schematic of a "TLUD" heating stove

Lobscheid et al. (2014) observed three main stove operating behaviors for TLUD stoves during a field survey. These include: 1) cold start in which the stove is empty and cold prior to loading and lighting fuel from the top (i.e., per manufacturers recommended operating procedure); 2) warm refueling in which cold fuel is added atop residual embers and ash in the combustion chamber. Residual heat being insufficient for immediate ignition, the fuel is lit from the top, but the pre-warmed stove could subsequently ignite from below; and 3) hot refueling where sufficient embers, flames, and heat is present in the combustion chamber at the time of refueling to ignite the fresh, cold fuel upon reloading with no top-lighting. The frequency of occurrence of each of these stove operating behaviors is discussed by Lobscheid et al. (2014). Emission factors for each scenario are needed to support real-world assessments of stove performance and to estimate stove emissions for exposure and health risk assessments. In addition, field-representative emission factors may help identify opportunities for additional improvements to stove design.

The combination of stove design, fuel characteristics, and operational factors, along with the complex nature of the combustion process, leads to significant levels of variability and uncertainty in emission measurements. Some of the variability is explained by differences in environmental and operational factors (field variability), but even more may be from measurement error (measurement uncertainty or measurement bias) in the field or in the laboratory.

Most of the Mongolian stove studies have relied on a measurement technique known as the "chimney method" in which emissions are sampled directly from the chimney using a custom dilution apparatus (Pemberton-Pigott, 2011). The chimney method involves a number of steps including sampling hot gas from the base of the chimney just above the fuel bed, applying a constant stream of clean dry dilution air at the inlet of a sample line, cooling of the gas stream as it travels from the sampling port to the analyzer(s), fog, water and soot condensation management in the inlet and sampling lines, and determination of undiluted chimney exhaust flows from fuel burn rates and mass balance calculations. Each of these steps can contribute to measurement uncertainty. Additionally, the chimney method "freezes" emissions sampled at the base of the chimney using cold dilution air. However, in a normal chimney, high temperatures, concentrated aerosols, and volatile gases lead to evolution of the aerosol inside the chimney; "frozen" aerosol measurements taken from the base of the chimney may differ from air quality-relevant emissions exiting the top of the chimney. Finally, although the chimney method is used extensively in Mongolian stove testing projects, there are virtually no published studies validating the robustness of this method.

In contrast, several published studies explain and support the use of a dilution tunnel approach in which samples are drawn from a calibrated dilution chamber mounted above the chimney. Advantages of the dilution chamber approach including (a) the method is focused on emission samples drawn from an ambient air quality-relevant location—the exit of the chimney and (b) the samples are diluted and measured in a large dilution tunnel rather than a small tube—an approach that virtually eliminates particle clogging and eases the sample dilution process. However, the dilution chamber approach is difficult to apply in the field because it requires accessing the top of the chimney.

A testing facility was designed and constructed at Lawrence Berkeley National Laboratory (LBNL) for measuring emissions from space heating stoves under simulated field-use conditions. It is important to understand differences between the sampling methods, therefore the LBNL test facility was instrumented with both the direct chimney dilution sampler and the top of chimney dilution tunnel. In this paper we report on the design of the LBNL test facility, the testing methods, and the results of emissions testing conducted on a Silver-mini (small Turkish) top-lit up-draft (TLUD) stove under several start-up and refuel scenarios. We evaluate the testing results in the context of three other studies of PM 2.5 emissions from heating stoves used in Mongolia. These studies differ according to test location (lab, field, or test Ger), measurement technique (chimney or exhaust dilution), and focus (fuel type, fueling behavior, and climate). We conclude with a discussion of the sources of variability and uncertainty in emissions testing and a summary of recommendations for future experiments to improve the estimation of actual emissions from improved space heating stoves.

2. Method

2.1 Experimental Facility

The LBNL test facility combines the direct flue-gas dilution approach (Iyer et.al, 2010; Pemberton-Pigott, 2011) and the dilution tunnel approach (Gullett, Touati & Hays, 2003; Purvis, McCrillis & Kariher (2000); Pettersson, Lindmark, Ohman, Nordin, Westerholm & Boman, 2010; Pettersson, Boman, Westerholm, Bostrom & Nordin, 2011; Boman, Nordin, Westerholm & Pettersson, 2005, Boman, Pettersson, Westerholm, Bostrom & Nordin 2011). The experimental facility, illustrated in Figure 2, was constructed inside a large high-ceilinged metal building. A brief description of the facility is provided below with more details in the Maddalena, Lunden, Wilson, Ceballos, Kirchstetter, Slack & Dale, 2014).

The test stove is placed on a platform scale with digital readout that records continuous mass during test burns. The continuous mass is used along with fuel composition to determine flue gas flow rate using a mass balance. The exhaust from the combustion chamber is vented to a section of chimney (0.127 m diameter × 0.610 m height) that contains the flue-gas dilution apparatus. A standard metal chimney (0.102 m diameter × 2.428 m height) is fit to the top of the flue-gas dilution apparatus. The dilution tunnel is suspended above the stove at 2.5 m height and the chimney extends 0.25 m into the dilution tunnel through a slightly enlarged hole. The chimney is free standing supported with cables to allow for continuous mass determination during a burn.

The dilution tunnel is constructed from a 1.83 m length of 0.61 m diameter duct suspended horizontally above the stove. The air inlet to the dilution tunnel on the right side of Figure 2 has an adjustable diameter to allow for control of the pressure in the dilution tunnel to adjust the chimney draft. The outlet from the dilution tunnel feeds into a 0.15 m diameter duct that passes through an adjustable damper to an industrial blower before exhausting from the building through a spark-arresting screen. Flow through the dilution tunnel is controlled and monitored by the iris damper and pressure inside the dilution tunnel is controlled by the size of the inlet.

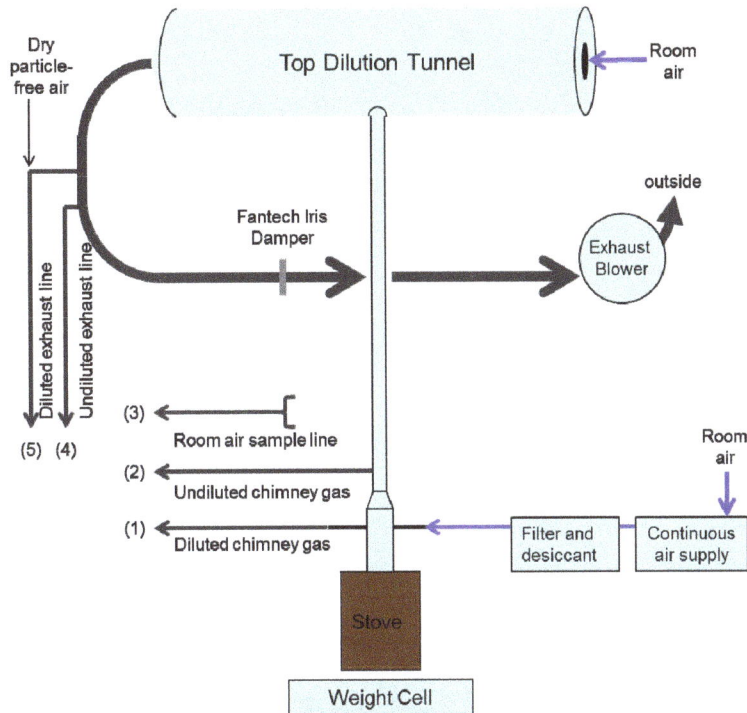

Figure 1. Schematic of LBNL Stove Testing Facility. All stove emissions and effluent from sample lines are exhausted outside. The flue-gas dilution apparatus is illustrated in Figure 2 and the numbered sampling lines are shown in detail in Figure 3

The chimney dilution apparatus is made up of two 12.7 mm heavy-walled stainless steel pipes mounted in adjustable sleeves. The sleeves are affixed on either side of the chimney, and the stainless steel pipes are allowed to slide so they meet near the center of the chimney. The tips of the pipes are milled so that the pipe delivering particle free dry air has a small orifice and the receiving pipe has an inverted cone shape. The pipes are mounted approximately 1 mm apart to allow chimney gas to be drawn into the sampling line. The difference in flow rate in the delivery line and the sample line are used to control the dilution. The design is based on that of the SEET lab (Pemberton-Pigott, 2011) where chimney gas is rapidly extracted from the chimney and diluted with dry particle free air to prevent moisture/aerosol condensation in the sample line and to bring gas concentrations within operational limits of the analyzers. The particle-free dry air is generated using a continuous flow compressor (Dewalt model D55146) and the air passes through a coalescing filter, two drying cartridges (Parker/Watt dryrite model DD15), a HEPA particle filter, and finally a mass flow controller for continuous metered flow (Alicat 0-20 LPM).

2.2 Instrumentation

There are five sample lines built into the system (see Figure 4) that measure diluted and undiluted gas from the chimney, diluted and undiluted gas from the dilution tunnel, and room air. In addition, the temperature is measured continuously in the room, inside the chimney just above the stove, inside the chimney at ceiling height (8 foot off floor), and in the 6-inch exhaust duct downstream from the upper dilution tunnel (Figure 2). Pressure is measured in the chimney just above the stove, in the chimney at the ceiling height, and in the exhaust duct on each side of the FanTech iris damper.

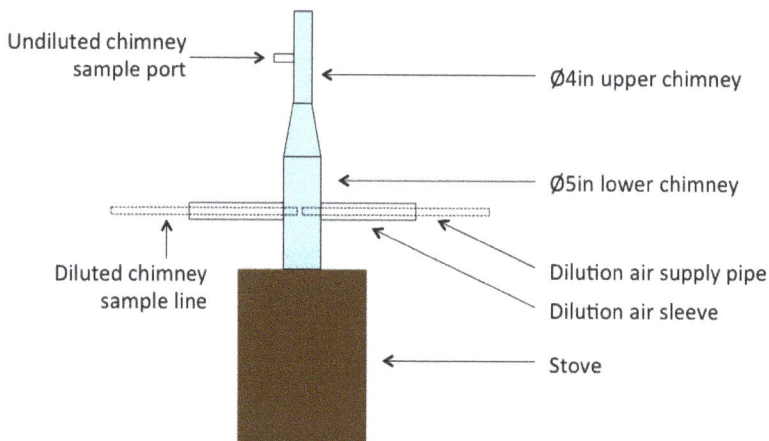

Figure 2. Detailed figure of the chimney dilution sampling system

Sample line 1 provides diluted chimney gas that passes through a 2.5 micron cyclone before being split to a gas analyzer (CAI 600 Series Model 602P – $CO_2/CO/O_2$), real time PM mass sensor (TSI DustTrak II model 8530), an integrated PM mass measurement (25 mm Teflon filters) and excess flow for additional lines to mount other instruments or gas samples as needed. Sample line 2 is undiluted chimney gas drawn through a coalescing filter filled with glass beads to reduce static volume followed by a Nafion drying column (MD-110-125-4) with dry air counter current flow (5 LPM) followed by a second gas analyzer (CAI 600 Series model 602P – $CO_2/CO/O_2$). Line 3 samples room air through a CO_2 analyzer (Li-Cor model LI-820). Sample line 4 is drawn directly from the exhaust after it exits the upper dilution tunnel and is sampled through a second CO_2 analyzer (Li-Cor model LI-820). The flow through both Li-Cor samplers is controlled by constant vacuum and critical orifice. Sample line 5 provides a secondary dry particle free source of air for additional dilution of the exhaust before running through a second real-time PM mass sensor (TSI DustTrak II model 8530).

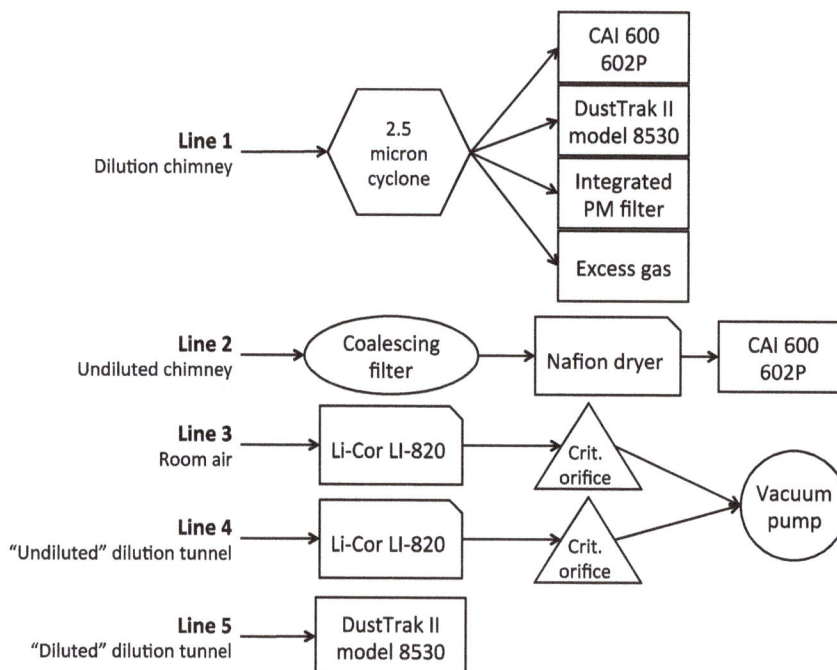

Figure 3. Sample lines and instruments

2.3 Emissions Testing

All emission tests were performed using a Silver-mini (small Turkish) TLUD stove and Nailakh coal, one of the most commonly used fuels in Ulaanbaatar (World Bank, 2011). The coal was shipped in sealed barrels from Mongolia to LBNL. The typical test included a cold start with approximately 10 kg of fuel followed by a refueling event consisting of approximately 5 kg fuel. The cold-start was accomplished by filling the fire-box with the specified amount of coal, placing a small amount of paper and dry wood on top of the coal and lighting the paper resulting in a "top lit down draft" condition. The refueling events were conducted at different stages of the burn with the earliest refueling event occurring as soon as the coal fuel bed collapsed and the latest refueling conducted while enough embers remained to ignite the coal. Some of the startup events used less than the 10 kg of coal to explore the impact on emissions during ignition. A summary of experiments is provided in Table 1. A stove conditioning burn was conducted prior to the first test to remove residues from the stove and to identify appropriate dilutions for the sampling lines.

Table 1. Summary of Experiments

Test Date	Test Name	Stages of burn	Description of test
20-March	Typical burn; early refuel	Cold-start Early hot-refuel	Typical mass of fuel at both ignitions Hot refueling performed just after fuel bed collapse
21-March	Typical burn; late large refuel	Cold-start Late hot-refuel	Typical mass of fuel at startup Extra fuel used at refueling Hot refueling performed late in run
27-March	Typical burn; late refuel	Cold-start Late hot-refuel	Typical mass of fuel at both ignitions Hot refueling performed late in run
28-March	Typical burn; no refuel	Cold start	Stove did not start initially. Typical mass of fuel used at startup No refuel event
13-June	Light burn; late refuel	Cold-start Late hot-refuel	Light load of fuel used at startup Typical mass of fuel used at refueling Refueling performed late in run Only exhaust dilution line used in PM sampling

2.3.1 Setting the Standard Chimney Draft

When conducting emission testing under laboratory conditions, it is important to set the chimney draft at a representative value for field conditions. The chimney draft (at ceiling height) for this study is set to be representative of winter conditions in Ulaanbaatar. We assume that the typical chimney consists of a 10 foot section of 4 inch pipe extended 1 meter above the roof line (ceiling height of a Ger). The air inlet to the stove is open at ~ 6 inches above the floor. Indoor and outdoor temperatures are 20°C and -20° C respectively, with a 5 meters-per-second wind speed at the top of the chimney.

The draft in this case is dominated by wind effect where the typical flue wind pressure coefficient is about -0.5 so the pressure gradient caused by wind (dP_{wind}) is estimated as

$$dP_{wind} = 0.5\rho * Cp * V^2 \qquad (1)$$

Where ρ is the density of air, C_p is the specific heat of chimney gases and V is wind speed at the top of the chimney. The resulting draft at ceiling height for wind effect is approximately - 7.5 Pa. For stack effect we assume 1.45 m of chimney is indoors and 1 m is outdoors. With no fire, the temperature of air in the flue is the same as the room temperature therefore the inside section of chimney generates no stack effect. The outside section of chimney generates a draft caused by the temperature gradient calculated as

$$dP_{stack} = \rho \times g \times h \times dT/T \qquad (2)$$

Where ρ is the density of air, g is acceleration of gravity (-9.8 m/s^2), h is the height of the section of chimney and dT/T is the temperature gradient relative to room temperature (K). The resulting draft for stack effect is approximately - 1.7 Pa. So the total draft ($dP_{wind} + dP_{stack}$) at ceiling height at startup is -7.5 - 1.7 = -9.2 Pa. We set the draft to a value between -7 and -10 Pa. The draft in the chimney is set by controlling flow in the dilution tunnel above the chimney.

3. Results

We found that virtually all of PM emissions (by mass) occur during ignition events (i.e., when the fuel is lit by the operator or by residual embers/flame in the fire box), while CO emissions were more widely distributed over different phases of the burn (Table 1). For the ignition events, we found that refueling produced up to five times more PM emissions than initial cold starts, with the one exception being an early refueling when the temperature in the fire-box was above 450 Celsius. For the cold starts, it appears that the majority of PM is caused by the ignition and combustion of wood and paper and only a moderate amount of PM is caused by the ignition of coal. Not surprisingly, for refueling events, all of the PM emissions result from the combustion of coal.

PM emission factors integrated over the entire burn and measured using the exhaust technique range from 3 – 8 grams per kg fuel consumed (Table 1). Emission factors for the ignition period range from 1 – 16 g/kg fuel consumed for cold-start conditions and from 7 – 81 g/kg fuel consumed for refuel events. The duration of ignition phases varied from run to run (range 21 – 89 minutes for cold starts and 20 – 41 minutes for refueling events.

While fuel consumption was measured directly during this study, measuring fuel consumption in the field during different phases of a burn is not feasible. Fortunately, particulate emissions result almost entirely from ignition events and are relatively consistent regardless of the amount of fuel or the duration of the ignition phase (for the conditions tested in this study). Therefore, we recommend reporting emissions per ignition event, which could greatly simplify the calculation of source terms for air quality modeling. As such, we found PM2.5 emission rates of 15 grams per cold start event (± 80% coefficient of variation) and 60 grams of PM2.5 per refuel event (± 80% coefficient of variation).

The large variance in results is due in part to the relatively small number of experiments and large number of factors that influence emissions. The variance could be reduced with experiments that focus on a small number of covariant factors. For example, we combined differences in operational behaviors with differences in fuel characteristics (size, moisture). We also varied the mass of coal loading and the amount of ash removed from the fire-box prior to refueling. The lack of experimental replicates limits our ability to provide a reliable estimate of uncertainty, but the largest contributor to uncertainty appears to arise from test imprecision—as suggested by the wide variation of emission estimates using the chimney technique.

Table 2. Summary of test data

Test Name and Type			Typical burn; early refuel	Typical burn; late large refuel	Typical burn; late refuel	Typical burn; no refuel	Light burn; late refuel
Metric	Test Type	Unit	20-Mar	21-Mar	27-Mar	28-Mar	13-Jun
Overview							
Test Duration		min	452	557	508	384	192
Initial Mass of Coal		kg	10.8	10.6	10.0	10.0	6.0
Initial Mass of Wood and Paper		kg	0.3	0.5	0.4	0.5	0.5
Refuel Mass of Coal		kg	4.7	6.8	5.0	0.0	4.9
Stove Temperature at Refuel		C	638	451	438	NA	364
Particulate Matter							
Total PM Emissions During Full Test	chimney	g	19.7	32.85	NA	3.19	NA
	exhaust	g	34.2	107	111	28.6	32.5
% of Total PM Emissions Occuring During Refuel	chimney	%	2.6%	63.7%	99.4%	NA	NA
	exhaust	%	25.3%	83.1%	95.2%	NA	96.5%
% of Total PM Emissions Occuring During Ignition Events	chimney	%	99.5%	99.7%	99.4%	98.1%	NA
	exhaust	%	98.5%	99.0%	99.2%	98.3%	99.1%
PM Emission Factor (PM emissions/fuel consumed) During Full Test	chimney	g/kg	1.45	2.30	NA	0.33	NA
	exhaust	g/kg	2.52	7.49	7.92	2.97	3.51
Carbon Monoxide							
Total CO Emissions During Test		g	439	339	282	326	145
% of Total CO Emissions Occuring During Startup		%	25.2%	35.2%	14.4%	20.6%	13.1%
% of Total CO Emissions Occuring During Refuel		%	10.9%	45.6%	57.0%	NA	44.7%

4. Discussion

In each LBNL emissions test, the chimney emission estimates are lower than the exhaust emission estimates. As noted previously, although the chimney method has been the most commonly used measurement technique for evaluating stove emissions, measurement errors associated with this technique may be large. We observed several anomalies with the chimney dilution sample line during the testing and the results are reported only for comparison and completeness. Several times we noted a significant drop in the measured concentration in the chimney dilution line. The concentration measured in the chimney dilution line converted to chimney concentration was almost always lower than the value measured with the exhaust dilution tunnel sample line. This occurred even after switching the DustTrak particle sampler used on the two lines. The gas phase measurements from the chimney dilution line did not indicate complete plugging of the line, but particles may have been lost to the walls of the collection cone in the chimney or the sample transfer line from chimney to instrument. In addition to errors caused by deposition of particles in the dilution apparatus, we also reiterate our concern about the practice of sampling aerosol at the base of the chimney before it has had time to fully evolve; chimney PM may be lower than dilution tunnel PM simply because aerosol did not have a chance to nucleate and grow before being "frozen" by the dilution apparatus.

A comparison of minute-by-minute PM estimates indicates the frequency with which the chimney method understated PM emissions compared to the exhaust method in our study (Figure 5). The bold 45-degree line indicates 1:1 data coorespondance where the exhaust measure of PM equals the chimney measure. Points below that line indicate exhaust estimates that exceeded the corresponding chimney estimates; points above the line indicate chimney estimates that exceeded the exhaust estimates. It is immediately apparent that the chimney estimates were lower than exhaust estimates in all but a few test observations. Of 1,886 emission estimates shown in the figure, all but 23 fall below the 1:1 line.

This tendency for the chimney method to understate PM emissions appears to be in rough proportion to the emission level. This is suggested by power regression curves relating chimney to exhaust emissions in four test runs (Figure 5). The regression curves slant below and away from the 1:1 line at higher emission levels. The regression coefficients suggest that the chimney method missed about 5% of the PM emissions in the early and large refuel runs and 17% of PM emissions in the late refuel run, relative to exhaust method estimates.

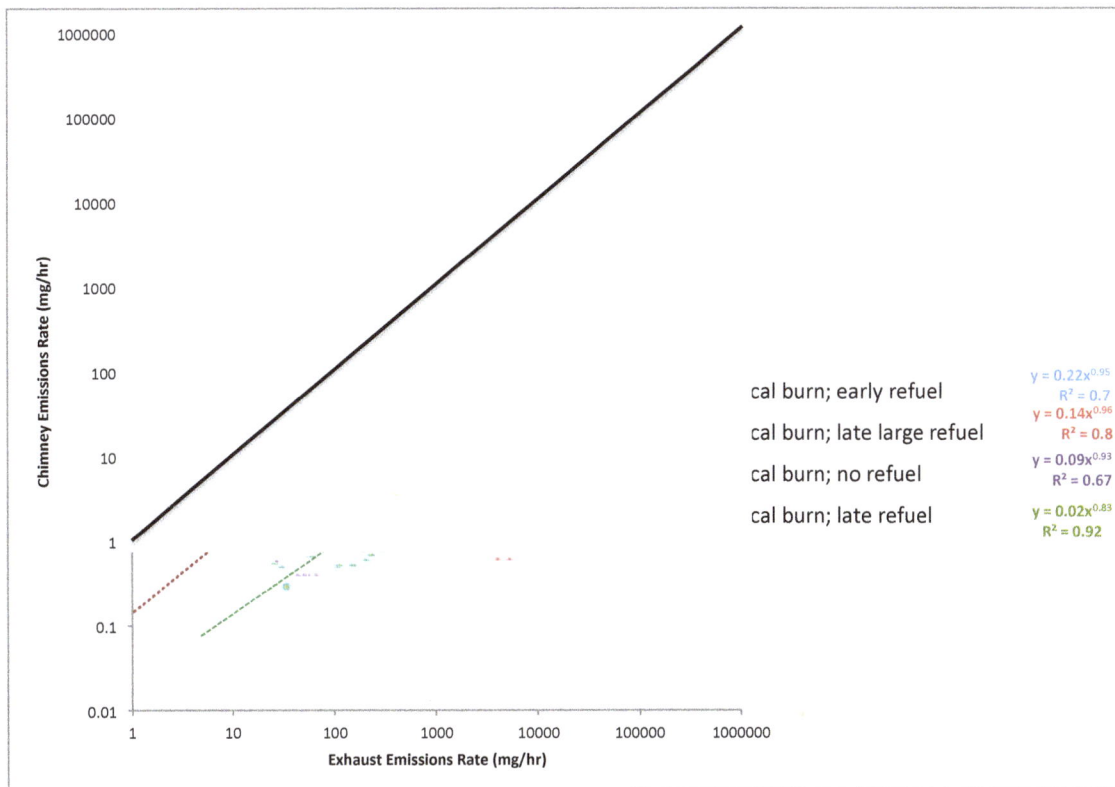

Figure 5. Comparison of exhaust and chimney emission detection

The chimney method emission estimates obtained in this study are consistent with the relatively low estimates reported in other studies using this direct chimney dilution method, including studies by the Building Energy Efficiency Center (BEEC) (Munkhbayar et al. 2011) and the Asian Development Bank (ADB) (ADB/

Pemberton-Pigott, 2010). The BEEC study estimated emissions from the Silver mini in different households over several fuel cycles including, presumably, late and early refueling. The ADB study measured Silver mini emissions in a laboratory setting without refueling of any type.

Obviously, emission estimates vary widely across these studies (Table 2). Differences in stove use practices explain much of the observed range. For example, emission estimates applying the chimney method range between .03 (g/Kg) without refueling and 2.3 (g/Kg) with late refueling. However, differences in measurement technique error may also explain some of the variation as illustrated in Figure 5. For example, emissions estimated using the chimney technique are generally lower, and sometimes much lower, than emissions measured using the exhaust technique.

Table 2. Total PM Emission Factor (g/Kg) --Selected Studies and Measurement Techniques

Studies	Study Type	Test Type	Unit	Emission Factor Estimates Small-Turkish Stoves		
				No Refuel	Early Refuel	Late Refuel
LBNL	Lab	Exhaust	g/kg	2.97	2.52	7.49
LBNL	Lab	Chimney	g/kg	0.33	1.45	2.30
BEEC	Lab	Chimney	g/kg		1.70	
ADB	Lab	Chimney	g/kg	0.03		

Source

LBNL   This study.

BEEC   Munkhbayar et. al. March 2011. "Results of Field Testing for Stoves, Gas and Electric Heater". Building Energy Efficiency Center." MCA Mongolia. Tbl 12.

ADB   ADB/Crispin Pemberton-Pigott. "Test 115." Stove Emissions and Efficiency Testing Laboratory. December 2010. Section 6.5. Pg 8.

The finding that much of the observed range of emission estimates may be due to measurement error and bias suggests, as a policy measure, that priority should be given to increasing the accuracy and relevance of the techniques used to measure emissions—either through improved calibration of existing measurement techniques or through development of new techniques.

Based on this evaluation of uncertainty and variability, we recommend additional laboratory study of the emissions from the Turkish mini and other approved stoves to confirm the accuracy of the exhaust technique in the lab and improve the accuracy of the chimney technique in the field.

Acknowledgments

The authors are extremely grateful to Crispin Pemberton-Pigott (New Dawn Engineering, Swaziland, Southern Africa), who generously shared from his experience testing and improving space heating stoves and details about the design of his direct chimney dilution apparatus. We thank the Gadgil Lab's Cookstove Research Facility (University of California, Berkeley (UCB) and LBNL) for sharing instrumentation and the capable team of students. We thank Iain Walker (LBNL) for calculations of the standard chimney draft used in this project and are grateful to Henrik Wallman (UCB and LBNL) for stimulating conversations, interesting anecdotes, and challenging questions.

This report was prepared as a result of work sponsored by the Millennium Challenge Corporation, U.S. under Federal Interagency Agreement MCC-10-0064-IAA-80, the California Energy Commission Public Interest Energy Research Program, Energy-Related Environmental Research Program, award number 500-09-049and by the U.S. Department of Energy under Contract No. DE-AC02-05CH11231.

Disclaimer

information, apparatus, product, or process disclosed, or represents that its use would not infringe privately owned rights. Reference herein to any specific commercial product, process, or service by its trade name, trademark, manufacturer, or otherwise, does not necessarily constitute or imply its endorsement, recommendation, or favoring by the United States Government or any agency thereof, or The Regents of the University of California. The views and opinions of authors expressed herein do not necessarily state or reflect those of the United States Government or any agency thereof, or The Regents of the University of California.

References

ADB/ Pemberton-Pigott C. (December 2010). "Test 115." Stove Emissions and Efficiency Testing Laboratory. Section 6.5, pg. 8.

Allen, R.W., Gombojav, E., Barkhasragchaa, B., Byambaa, T., Lkhasuren, O., Amram, O., Takaro, T. K., & Janes, C. R. (2013). "An assessment of air pollution and its attributable mortality in Ulaanbaatar, Mongolia" Air Quality. *Atmosphere & Health, 6*(1), 137-150. http://dx.doi.org/10.1007/s11869-011-0154-3

Boman, C., Nordin, A., Westerholm, R., & Pettersson, E. (2005). "Evaluation of a constant volume sampling setup for residential biomass fired appliances – influence of dilution conditions on particulate and PAH emissions". *Biomass and Bioenergy, 29*, 258-268. http://dx.doi.org/10.1016/j.biombioe.2005.03.003

Boman, C., Pettersson, E., Westerholm, R., Bostrom, D., & Nordin, A. (2011). "Stove Performance and Emission Characteristics in Residential Wood Log and Pellet Combustion, Part 1: Pellet Stoves". *Energy Fuels, 25*, 307-314. http://dx.doi.org/10.1021/ef100774x

Davy, P. K., Gunchin, G., Markwitz, A., Trompetter, W. J., Barry, B. J., Shagjjamba, D., & Lodoysamba, S. (2011). "Air particulate matter pollution in Ulaanbaatar, Mongolia: determination of composition, source contributions and source locations". *Atmospheric Pollution Research, 2*, 126-137. http://dx.doi.org/10.5094/APR.2011.017

Gullett, B. K., Touati, A., & Hays, M. D. (2003). "PCDD/F, PCB, HxCBz, PAH, and PM Emission Factors for Fireplace and Woodstove Combustion in the San Francisco Bay Region". *ES&T, 37*(9), 1758-1765. http://dx.doi.org/10.1021/es026373c

Iyer, M., Wallman, H., & Gadgil, A. (Draft October 2010). Strategies for Reducing Particulate Emissions from Space Heating in the Ger District of Ulaanbaatar, Mongolia. Lawrence Berkeley National Laboratory Report Number LBNL-Draft.

Lobscheid, A., Fitts, G., Lodoysamba, S., Maddalena, R., & Dale, L. (2014). "Pilot Study of Fuel and Stove Use Behavior of Mongolian Ger Households". Lawrence Berkeley National Laboratory Report LBNL-6543E Retrieved from http://escholarship.org/uc/item/9sr4r2mf#page-1

Lodoyasamba, S., & Pemberton, P. C. (2011). Mitigation of Ulaanbaatar city's air pollution – from source apportionment to ultra-low emission lignite burning stoves (pp. 27-30). Proceedings of the 19th International Conference Domestic Use of Energy, 12-13 April, Cape Town. ISBN 978-0-9814311-4-7.

Maddalena, R, Lunden, M., Wilson, D., Ceballos, C., Kirchstetter, T., Slack, J., & Dale, L. (2013). "Quantifying Stove Emissions Related to Different Use Patterns for the Silver-mini (Small Turkish) Space Heating Stove" Lawrence Berkeley National Laboratory. October, 2013, Report LBNL-6319E. Retrieved from http://eetd.lbl.gov/publications/quantifying-stove-emissions-related-t

Munkhbayar et al. (March 2011). "Results of Field Testing for Stoves, Gas and Electric Heater." Building Energy Efficiency Center. MCA Mongolia. Tbl 11.

Pemberton-Pigott C. (2011). Development of a low smoke Mongolian coal stove using a heterogeneous testing protocol. Proceedings of the 19th International Conference Domestic Use of Energy, 12-13 April, Cape Town. ISBN 978-0-9814311-4-7, pp. 65-70.

Pettersson, E., Boman, C., Westerholm, R., Bostrom, D., & Nordin, A. (2011). "Stove Performance and Emission Characteristics in Residential Wood Log and Pellet Combustion, Part 2: Wood Stoves". *Energy Fuels, 25*, 315-323.

Pettersson, E., Lindmark, F., Ohman, M., Nordin, A., Westerholm, R., & Boman, C. (2010). "Design changes in a fixed-bed pellet combustion device: effects of temperature and residence time on emission performance" *Energy Fuels, 24*, 1333-1340.

Purvis, C. R., McCrillis, R. C., & Kariher, P. H. (2000). "Fine Particulate Matter (PM) and Organic Speciation of Fireplace Emissions". *ES&T, 34*(9). http://dx.doi.org/10.1021/es981006f

World Bank. (2011). Main report. Vol. 1 of Air quality analysis of Ulaanbaatar: improving air quality to reduce health   impacts.   Washington,   DC: World   Bank.   Retrieved   from http://documents.worldbank.org/curated/en/2011/12/15633946/air-quality-analysis-ulaanbaatar-improving-air-quality-reduce-health-impacts-vol-1-2-main-report

13

Carbon Based Electrochemical Double Layer Capacitors of Low Internal Resistance

Yurii Maletin[1,2], Volodymyr Strelko[2], Natalia Stryzhakova[1,2], Sergey Zelinsky[1,2], Alexander B. Rozhenko[1], Denis Gromadsky[1], Vitaliy Volkov[3], Sergey Tychina[1,2], Oleg Gozhenko[1,2] & Dmitry Drobny[1,2]

[1] YUNASKO-Ukraine, Kiev, Ukraine

[2] Institute for Sorption and Problems of Endoecology, National Academy of Science of Ukraine, Kiev, Ukraine

[3] Institute of Chemical Physics, Russian Academy of Science, Chernogolovka, Moscow Region, Russia

Correspondence: Yurii Maletin, YUNASKO-Ukraine, Kiev, Ukraine. E-mail: ymaletin@yunasko.com

Abstract

Based on nanoporous carbon electrodes electrochemical double layer capacitors (EDLC), otherwise known as supercapacitors or ultracapacitors, are currently widely used in various energy storage technologies, wherein the EDLC low internal resistance and long cycle life are at an advantage. It is still a good challenge to further reduce the internal resistance of EDLC since this can result in higher power density and higher efficiency of these promising power supply units. In this work it has been found that the EDLC internal resistance depends strongly on the electrolyte diffusion in the carbon electrode nanopores, and two techniques to measure the in-pore diffusion coefficients, namely, those based on spin-echo NMR or cyclic voltammetry with the use of porous rotating disc electrode are described. Cyclic voltammetry, impedance spectroscopy and transmission electron microscopy have also been used to select the best EDLC components. As a result, EDLC devices of very low internal resistance and high power density have been developed.

Keywords: double layer capacitor, in-pore electrolyte diffusion, low resistance

1. Introduction

1.1 Why It Is of Interest

There is an obvious increasing interest in electrochemical double layer capacitor (EDLC) technology and application all over the world, in particular, in renewable energy and hybrid vehicle applications. However, the EDLC market growth is still rather modest. In our opinion, the main reason for a certain skepticism from automakers side is, on the one hand, a fast progress in Li-ion technology over the past two decades, while, on the other hand, a rather slow progress in EDLC technology. As pure "physical" devices, which do not involve any chemical or electrochemical transformations, any charge or mass transfer through the electrode-electrolyte interface, EDLC's must demonstrate much faster charge/discharge operations and longer cycle life than any "chemical" batteries (Conway, 1999). Given this, EDLC devices can provide the key to a number of efficient power solutions that are mainly related with various backup systems to compensate short-term voltage surges or drops or with load leveling the batteries in various combined power sources. Low internal resistance can be one of key advantages of EDLC over all other types of energy storage devices since the round trip efficiency and power capability of the devices are inversely proportional to their internal resistance. Besides, low internal resistance predetermines the effectiveness of EDLC application in combined power supply units, wherein an EDLC device is connected with a battery either in parallel or in series. Additionally, high efficiency implies low heat generation and, hence, improved safety. This, accompanied by EDLC long life cycle and wide operation temperature range, can help them to clear their way to the market. To realize it, EDLC devices must clearly demonstrate much higher power capability than Li-ion or any other type of batteries, but this is not always the case. What are the reasons?

1.2 What Are the Hurdles to Overcome

The most substantial contributors to the EDLC internal resistance are (see a simplified equivalent circuit in sketch (1) below): contact resistance between Al current collector and active carbon electrode (R_{Al-C}); ohmic resistance of active carbon electrode layer (R_C); electrolyte resistance in the electrode nanopores ($R_{El-in-pores}$); and electrolyte resistance in the bulk solution including electrode and separator macropores ($R_{El-in-bulk}$).

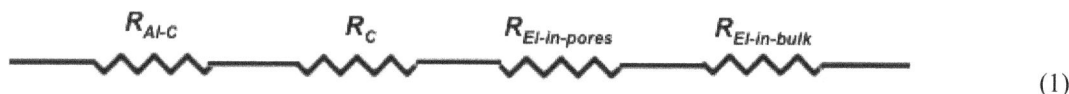

$$R_{Al-C} \quad\quad R_C \quad\quad R_{El-in-pores} \quad\quad R_{El-in-bulk}$$

$$\text{——}\diagdown\!\!\diagup\!\!\diagdown\!\!\diagup\text{——}\diagdown\!\!\diagup\!\!\diagdown\!\!\diagup\text{——}\diagdown\!\!\diagup\!\!\diagdown\!\!\diagup\text{——}\diagdown\!\!\diagup\!\!\diagdown\!\!\diagup\text{——}$$

(1)

High contact resistance, R_{Al-C}, can result from the native oxide film on the aluminum surface and its effect on EDLC performance is thoroughly discussed in our previous work (Maletin et al., 2008) and also in Section 4 below.

Ohmic resistance of the active electrode layer, R_C, can substantially be reduced by adding a small amount (2–5% wt.) of conductive additives such as carbon black or graphite. On the other hand, it has been found (Maletin et al., 2008; Maletin et al., 2012) that the electrolyte conductivity, though being high enough in the bulk solution, can significantly be reduced in the electrode nanopores, and this can result in the lion's share of an unexpectedly high internal resistance of EDLC devices. Experimental study of this phenomenon is the main subject of the present paper.

1.3 Our Approach to Solve the Problem

According to well-known equations (Bard & Faulkner, 2001, p.137) the electrolyte conductivity is proportional to mobility or to diffusion coefficients of the corresponding ions. On the other hand, there is no potential gradient in narrow pores (if the pore width is close to the Debye length, or to 1–2 nm in concentrated electrolytes, which are normally used in EDLC technology), and therefore, diffusion is the only driving force for charge-discharge processes in nanoporous EDLC electrodes (Maletin et al., 2006). Bearing that in mind, in the present study two known independent experimental techniques based on NMR or electrochemical measurements have been modified and used to measure the diffusion coefficients of the electrolyte inside the carbon nanopores.

Since cost is another important issue that hinders the EDLC way to the market, in the present study we are mostly focused on the low cost nanoporous carbons based on natural carbonaceous materials, e.g., coconut shell. In some cases the surface of initial carbons was doped with N-heteroatoms as was offered by Strelko, Stryzhakova, Gozhenko, and Maletin (2009).

2. Method

2.1 Design of EDLC Prototypes and Their Performance Measurements

Two-electrode capacitor prototypes were used for performance measurements of various nanoporous carbons and EDLC design solutions. The electrodes were typically prepared by mixing the nanoporous carbon powder with PTFE suspension in water (the latter was used as a binder) until a homogeneous mixture was obtained. No conductive additives were normally added since the carbons selected in this work provided fairly low ohmic resistance, R_C. The mixture was rolled to form sheets of 40–100 micron thick followed by cutting off the separate carbon electrodes. The active carbon electrodes thus obtained had their geometric surface area of 15 cm^2 each and contained 93% of carbon and 7% of PTFE binder. They were then applied onto electric-spark treated aluminum foil (Maletin et al., 2008) used as a current collector of 15 or 20 micron thick and dried at 220 °C under vacuum for 6 hours. A couple of electrodes were then interleaved with a porous insulating sheet (separator) and placed into laminated aluminum shell. The prototypes thus fabricated were filled with 1.3 M Et$_3$MeN (TEMA) BF$_4$ in acetonitrile and sealed. All the assembly operations were carried out in a dry box. Larger EDLC devices comprising a number of positive and negative electrodes connected in parallel and forming a stack with the capacitance between 400 and 1500 F were assembled at Yunasko Pilot Plant according to the same technology.

Cyclic voltammetry (CV) and electrochemical impedance spectroscopy (EIS) measurements were carried out with the help of Voltalab-80 PGZ-402 unit. Galvanostatic charge-discharge cycling with the help of Arbin BT-2000 testing unit was also used to measure the capacitance and internal resistance of large EDLC prototypes. The CV measurements were mostly carried out within the voltage range of 0–3 V with the scan rate of 10 mV.s^{-1}. In some cases the voltage range was extended to 3.5 V. Also in some cases three-electrode CV measurements were used to study the behavior of various EDLC components in either positive or negative voltage range. The current loads between 0.1 and 1.0 A.F^{-1} were used for galvanostatic charge-discharge cycling of the prototypes of different size/capacitance with the sampling rate of 10 ms. All the measurements were carried out at 25 °C except special life cycle tests that were carried out at 60 °C.

The internal resistance (R_{in}) and capacitance (C) values were calculated from galvanostatic cycling results in accordance with the FreedomCAR Ultracapacitor Test Manual (2004).

2.2 Study of Nanoporous Carbon Structure and Surface Chemistry

The following nanoporous carbon powders have been chosen for this study:

A. ZL-302 (Huzhou Sensheng Activated Carbon Co., Ltd);
B. ZL-302-N (ZL-302 carbon powder thermally treated with melamine and containing 15 atomic % of nitrogen on its surface);
C. NY1151 (Kuraray Chemical Co.,Ltd);
D. YP80F (Kuraray Chemical Co.,Ltd);
E. YP50F (Kuraray Chemical Co.,Ltd);
F. HDLC 20B STUW (Haycarb PLC);
G. NC2-1E (EnerG2 Technologies Inc.);
H. P2-15 (EnerG2 Technologies Inc.);
I. Y-Carbon (Y-Carbon Inc.).

The porous structure of carbon materials has been studied with the help of transmission electron microscopy (TEM) with the use of Jeol JEM-2100F, and also from nitrogen adsorption/desorption isotherms at 77 K using a Nova 2200e Surface Area & Pore Size Analyser (Quantachrome Instruments).

The concentration of nitrogen heteroatoms on carbon surface has been measured with the help of XPS with the use of KRATOS-800XPS, energy resolution of 1.2 eV, Kα (Al), hν=1486.6 eV, spectral data being treated with the help of XPSPeak 4.0.

2.3 Diffusion Coefficient Measurements

For NMR diffusion measurements the carbon powders listed above were impregnated with ethyl trimethyl ammonium tetrafluoroborate (EtMe$_3$NBF$_4$) dissolved in either acetonitrile (CH$_3$CN) or acetonitrile-d_3 (CD$_3$CN) followed by placing each powder into a 5 mm standard NMR tube. All ^{1}H NMR diffusion measurements have been carried out using a Bruker AVANCE 400 spectrometer equipped with the wide-bore magnet. A 5 mm "diff60" diffusion probe head (Bruker) with gradient coils in Z direction was used to generate magnetic field gradient. All experiments were performed at 25 °C. The 90° pulse lengths (14.5–18.5 µs) were determined for every sample using the standard routine. The T$_1$ relaxation times were measured by standard inversion-recovery. The T$_2$ relaxation times were determined with the Carr-Purcell-Meiboom-Gill (CPMG) standard sequence. The PGSTE "Pulse Gradient Stimulated Echo" method (DifSte Bruker standard pulse sequence) described by Tanner (1970) and by Cohen, Avram, & Frish (2005) was used for the diffusion measurements. The "diff" automated routine was employed to prepare all the parameters for the diffusion experiments. The square field gradient pulses (δ) and delay between two first radiofrequency pulses (τ) were chosen short enough (0.4 ms and 1.1 ms, respectively) in order to measure the diffusion coefficients for samples with very short (< 2 ms) T$_2$ relaxation times. 32 intensity points were acquired and number of transitions was between 8 and 64, depending on the signal intensity and resulting signal-to-noise ratio. The data were processed in the standard way using the T$_1$/T$_2$ routine and approximating the resulting curve on two different values of diffusion coefficients. The resulting values of self-diffusion coefficients were averaged over three measurements.

The ^{19}F NMR diffusion measurements have been carried out with the use of AVANCE-III-400 spectrometer following the same procedure as described above for ^{1}H self-diffusion experiments.

As an alternative independent method for diffusion coefficient measurements, a version of the well-known rotating disc electrode (RDE) method (Bard & Faulkner, 2001, p.335) has also been used. The disc electrode (BASi RDE-1) was made of a graphite rod of 3 mm in diameter and covered with a nanoporous carbon layer of 40 micron thick. Rotation rates of 500, 750, 1000, 1250 and 1500 rpm were chosen for measurements. As a reversible redox-pair for RDE measurements, the ferrocene molecule/ferrocenium cation (Fc/Fc$^+$) pair has been chosen since the Fc$^+$ cation is similar by its size and insignificant solvation effect to tetraalkylammonium cations used in EDLC technology. For RDE diffusion current measurements an Fc sample of 4 mmol.dm^{-3} was dissolved in 1.3 M Et$_3$MeNBF$_4$ solution in acetonitrile. The potential scan rate in CV measurements was 0.5 mV·s^{-1}. The Ag,AgCl/KCl (1 M) reference electrode was connected with the working electrolyte through a salt bridge.

3. Results

3.1 Study of the Carbon Pore Structure

Most of the carbon materials listed in Section 2.2 have been selected so that they have their pore size in the range between 1 and 10 nm with a large portion concentrated between 1 and 3 nm. Some examples can be seen in Figure 1 wherein the pore size distribution for three promising carbons is illustrated.

One more criterion for preliminary carbon material selection in this work is the presence of mostly shallow slit-shaped pores or just shear cracks of graphene layers in the carbon matrix – see, e.g., in Figure 2, which presents the TEM image for HDLC 20B STUW (F) carbon powder.

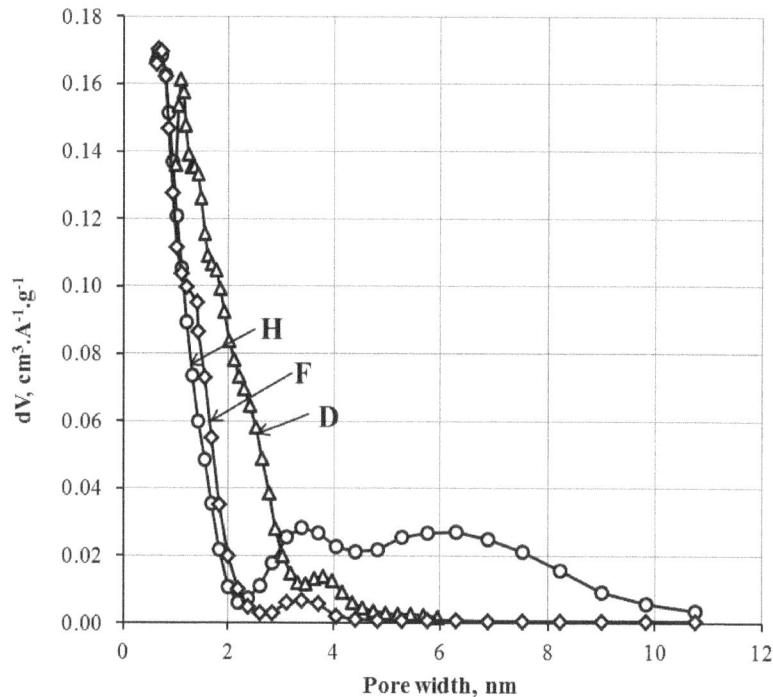

Figure 1. Pore size distribution obtained from nitrogen adsorption/desorption isotherms with the use of DFT calculations for some of the tested carbon powders (as listed in Section 2.2)

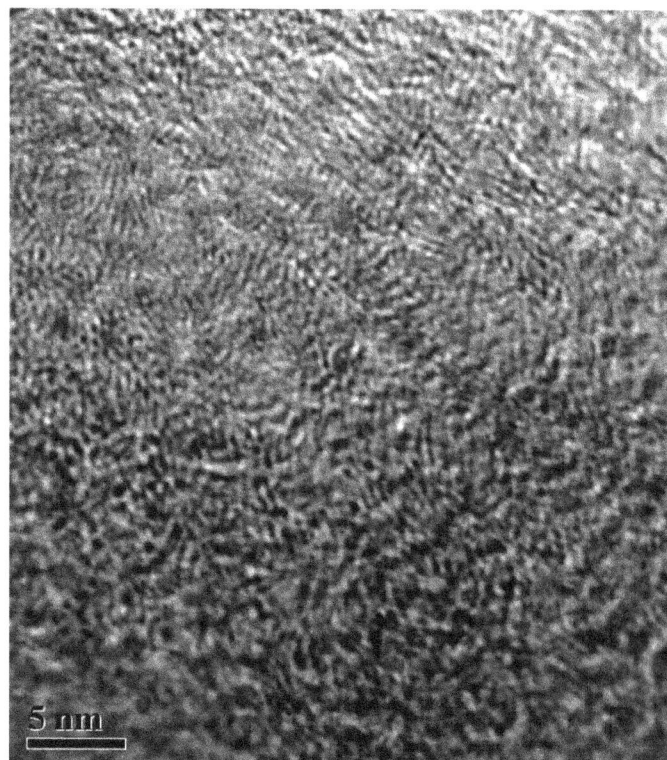

Figure 2. TEM image of F (HDLC 20B STUW) carbon powder

3.2 Electrochemical Measurements

Further, the carbon materials listed in Section 2.2 were studied with the use of CV method. A special reference electrode was developed to be used in organic electrolytes, the electrode potential being stable over practically unlimited time and giving a chance to register CV and charge-discharge curves in both positive and negative ranges from the equilibrium potential of carbon material. For the sake of generality, we use the Li/Li$^+$ potential as "zero point" in our scale of potentials.

A typical example of CV measurements with the use of three-electrode cell is illustrated in Figure 3-a, wherein a good behaviour of carbon materials selected for positive or negative electrode can be seen in a wide potential range. It should also be noted that a carbon material selected for the positive electrode may be different from that selected for the negative electrode to keep the electrode potential within the certain range when assembling the EDLC. As a result, a CV curve of an EDLC prototype assembled with the use of thus selected electrodes is presented in Figure 3-b.

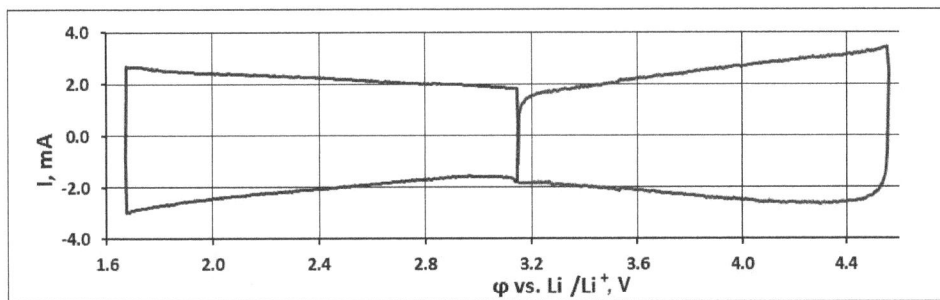

Figure 3-a. Cyclic voltammery curves in positive (right) and negative (left) ranges from the equilibrium potential of nanoporous carbon as a working electrode in a three-electrode cell (room temperature; electrolyte: 1.3 M TEMA BF$_4$ in acetonitrile; scan rate: 10 mV·s^{-1})

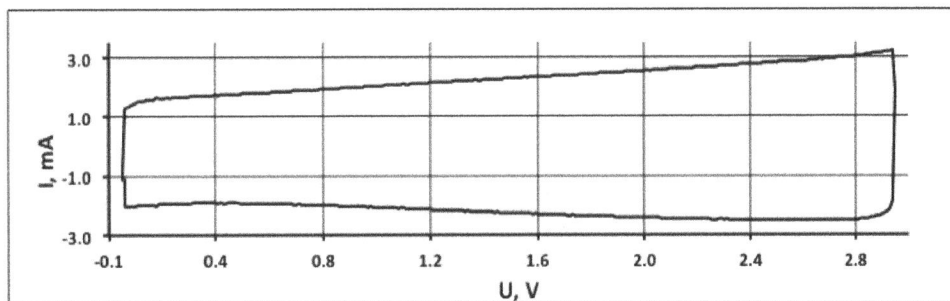

Figure 3-b. Cyclic voltammery curve of an EDLC prototype with positive and negative electrodes as in Figure 3-a, each electrode area being of 15 cm^2 (room temperature; electrolyte: 1.3 M TEMA BF$_4$ in acetonitrile; scan rate: 10 mV·s^{-1})

Figure 4 illustrates the EIS results (Nyquist plots) for three different designs of EDLC devices. Curve 1 illustrates the case of poor electrical contact between the aluminium current collector and active electrode layer resulting in high contact resistance. Curve 2 reflects the design with low contact resistance though with the electrochemical system (i.e. nanoporous carbon electrodes and organic electrolyte) non-optimized properly. The optimization can be achieved due to CV measurements like those presented in Figure 3 above and also due to diffusion coefficient measurements described below. The Nyquist plot for the optimized design is presented by Curve 3 in Figure 4.

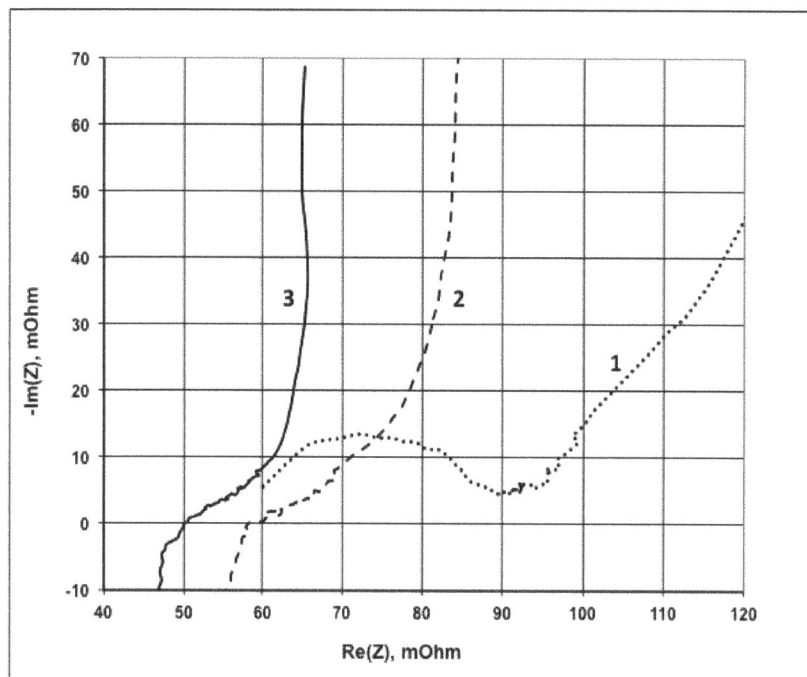

Figure 4. Nyquist plots for EDLC prototypes: 1 - with poor contact between the current collector and active carbon layer (high contact resistance); 2 - with improved contact resistance but non-optimized electrochemical system; 3 - with fully optimized design

3.3 In-Pore Diffusion Coefficient Measurements

After impregnating carbon powders with electrolyte, the electrolyte translational self-diffusion coefficients have been calculated from the attenuation of the NMR signals due to the application of the gradient pulses of the constant lengths but various strengths using Equation (2) (Price, 1997):

$$I = I_0 \, exp\left[-D_0\gamma^2\delta^2 g^2\left(\Delta - \frac{\delta}{3} \right) \right]$$

(2)

where I and I_0 are the NMR signal intensities in the presence and in the absence of the gradient respectively, γ is the gyromagnetic ratio of the nucleus under observation, δ and g are the duration and the strength of the applied gradient pulse, respectively, D_0 is the self-diffusion coefficient, and Δ is the time interval between the two successive gradient pulses.

It has been found that the theoretical curve practically coincides with experimental values if the attenuation of NMR signals is expressed as a sum of two Equations (2) with different D_0 values, a quickly diffusing one ($\sim 10^{-9}$ m^2 s^{-1}) and slowly diffusing one ($\sim 10^{-10}$ m^2 s^{-1}). This can reflect the fast diffusion in bulk solution or in macropores and slow diffusion in nanopores.

The resulting values of thus found effective diffusion coefficients have been averaged over three measurements and are plotted in Figures 5 and 6 (for BF_4^- anions and $EtMe_3N^+$ cations, respectively) for various nanoporous carbons vs. the internal resistance of EDLC prototypes comprising the same carbons in electrodes. As can be seen from these figures, there is a fairly good correlation between diffusion coefficients of electrolyte ions in carbon nanopores and the internal resistance of the corresponding EDLC prototypes. However, it should be noted that some carbons cannot be used for NMR measurements because of the very short correlation time of electrolytes in their pores resulting in unreliable diffusion coefficient evaluations. Therefore, we have also developed another method to measure the diffusion coefficients of electrolyte ions in carbon nanopores, namely, a version of the well-known RDE method-see Section 2.3 for details. The behaviour of porous rotating disc electrode (PRDE) has recently been studied by Bonnecaze, Mano, Nam, and Heller (2007). To evaluate the diffusion coefficients from PRDE measurements we have used the Levich equation:

$$i = \pm 0.62 n C_0 F D^{2/3} v^{-1/6} \omega^{1/2}$$

(3)

where i is the limiting diffusion current value, n is the number of electrons in the electrode semi-reaction, C_0 is the concentration of the reacting species, F is the Faraday constant, D is the diffusion coefficient, v is the kinematic viscosity, and ω is the rotation rate.

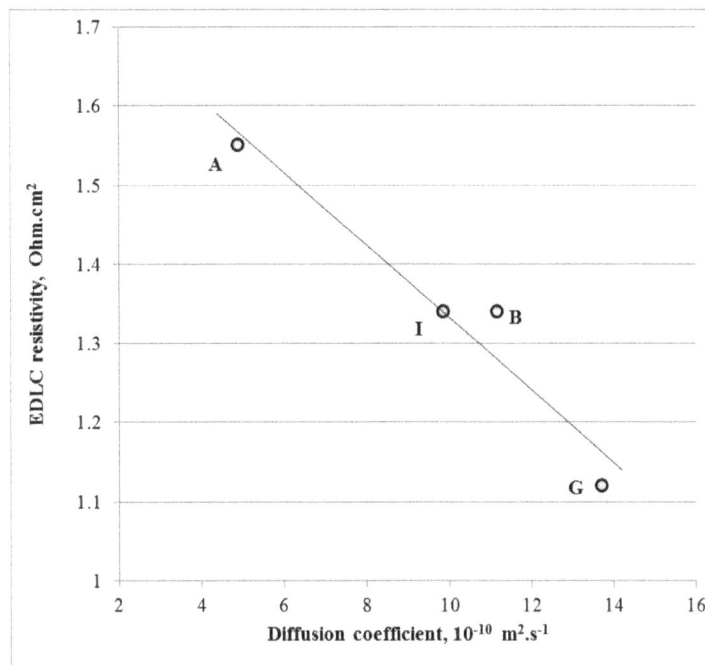

Figure 5. Internal resistance of EDLC prototypes comprising electrodes of different nanoporous carbons, as listed in Section 2.2, vs. the diffusion coefficients of BF_4^- anions in those carbons (^{19}F NMR measurements)

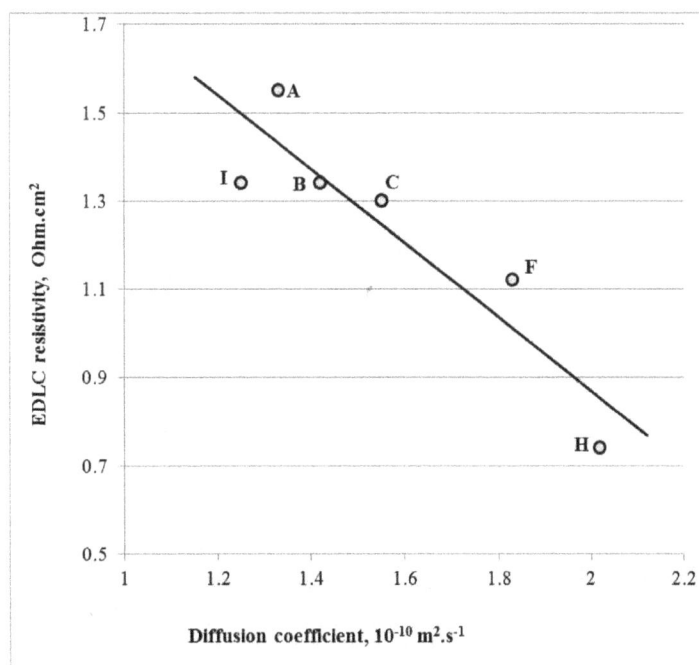

Figure 6. Internal resistance of EDLC prototypes comprising electrodes of different nanoporous carbons, as listed in Section 2.2, vs. the diffusion coefficients of $EtMe_3N^+$ cations in those carbons (^1H NMR measurements)

The results of PRDE measurements of the electrolyte diffusion in various nanoporous carbons, which are listed in Section 2.2, are plotted in Figure 7 vs. the internal resistance of EDLC prototypes comprising the same carbons in electrodes.

Both NMR and PRDE methods, as illustrated in Figures 5–7, have been employed to select the low resistance electrodes for EDLC manufacture, and the results for large EDLC prototypes are briefly listed in Table 1 below.

Table 1. Performance of Yunasko EDLC devices (rated voltage 2.7 V)

Capacitance, F	Internal resistance, mOhm	Time constant (RC), s	Spec. energy $(CU^2/2)$, W.h kg^{-1}	Spec. power (@95% eff.), kW kg^{-1}	Max. spec. power, kW kg^{-1}
480[a]	0.20	0.10	4.9	10.2	91
1200[a,b]	0.10	0.12	5.3	8.9	79
1500[b]	0.09	0.14	6.1	9.1	81

[a] Also tested at the Institute of Transportation Studies, Davis, CA; [b] Also tested at JME Inc., Cleveland, OH.

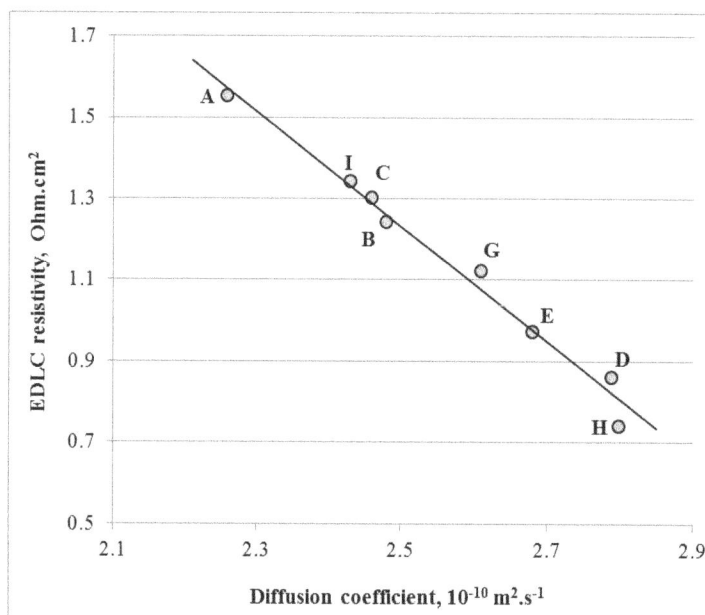

Figure 7. Correlation between EDLC internal resistance and diffusion coefficients of ferrocenium-cation in pores of various carbons listed in Section 2.2 and used in electrodes (PRDE measurements)

The cells listed in Table 1 were also tested for their life cycle according to IEC 62391 endurance test procedure, and after holding the cells for at least 2000 hours at 60 °C and 2.7 V they demonstrated the capacitance retention within 70% and an increase in internal resistance not exceeding 100%.

4. Discussion

Though being beyond the scope of this paper, some methods to reduce the R_{Al-C} contact resistance (see sketch (1)) have recently been disclosed by Maletin et al. (2008). The electrical contact between the active electrode layer and aluminum current collector can be significantly improved due to conductive carbon particles to be locally and individually fused into the current collector surface. New versions of those methods, in particular, the electric spark treatment with the use of graphite electrode are effectively employed in our current technology resulting in a very low contact resistance value, normally not exceeding 10 mOhm *cm^2.

As regards the choice of nanoporous carbon materials to be used in EDLC electrodes, the materials having shallow slit-shaped pores with their width between 1 and 3 nm (see examples in Figures 1 and 2) look preferable for EDLC electrode application. From our experience the carbons comprising larger mesopores have significantly lower surface area while those with pores predominantly below 1 nm demonstrate too high resistance to be effectively used in EDLC devices.

CV measurements with the use of a three-electrode cell and EIS measurements give a chance to optimize the electrochemical system design in order to improve the EDLC performance. Similar CV and dilatometric studies for different carbon materials and electrolytes were carried out by Hahn, Barbieri, Gallay, & Kötz (2006), and as have now been shown in our work the optimized EDLC prototypes demonstrate a fairly good rectangular shape of CV curves with no visible faradaic processes within a voltage range as wide as 2.9 V and also the lowest resistance and close to ideal vertical line in Nyquist plots at low frequencies– see Figures 3 and 4.

As was found in our NMR measurements, the T_1 and T_2 relaxation times of various nuclei of the electrolyte species (^1H, ^{11}B, ^{19}F) change significantly when the electrolyte enters the carbon pores, and this can reflect a strong interaction of electrolyte species with the carbon matrix – see also a similar conclusion made by Price (1997). This phenomenon and some other NMR results will be discussed in more detail in our forthcoming publications, while here we would like to focus on the diffusion coefficient values since they illustrate the electrolyte dynamics in carbon nanopores. Of course, the D_0 values measured for the fluid phase in nanopores are not the true self-diffusion coefficients, but rather the effective molecular diffusivities D_{eff} (Price, 1997). Nevertheless, as can be seen from Figures 5–7, there is a remarkable correlation between the internal resistance of EDLC device and ion diffusion in pores of the corresponding carbon electrode material. Absolute values of diffusion coefficients of $EtMe_3N^+$ and Fc^+ cations, though being measured by different methods, are close (cf. Figures 6 and 7), and the difference can be accounted for slightly different size of these cations. It is also worth noting that the diffusion coefficient of Fc^+ cations, when being measured with the use of a flat graphite electrode, is 1.01×10^{-9} $m^2 s^{-1}$, which significantly exceeds the values in carbon nanopores presented in Figure 7.

It should also be noted that the positive and negative electrodes are typically different to best match the different mobility of anions and cations in their pores. Besides, it has been found that the diffusion coefficients of electrolytes can be increased and, correspondingly, the pore resistance can be reduced due to doping the carbon surface with nitrogen atoms (Strelko et al., 2009). This effect can be accounted for reducing the interaction between the carbon matrix and electrolyte ions.

The test results listed in Table 1 and also verified by independent experts clearly show a very low internal resistance of EDLC devices based on optimized electrochemical systems with their RC-constant values being between 0.1 and 0.2 s and maximum power density reaching 80–90 kW/kg. These RC-constant values are lower by a factor of 2–3, and correspondingly, power capabilities of EDLC devices presented in Table 1 are by far higher than those of the best competing devices (Miller, Butler, Ryan, & McNeal, 2013).

Very low internal resistance of EDLC devices thus fabricated makes it possible to avoid significant heating during the operation, even under high load conditions. As an example, a module of 16 V comprising 6 single cells of 1200 F each connected in series has the internal resistance below 1 mOhm and after continuous charge/discharge cycling with the current value of 200 A running over 8 hours demonstrates an increase in temperature of about 10–12 °C only.

References

Bard, A. J., & Faulkner, L. R. (2001). *Electrochemical Methods. Fundamentals and Applications* (2nd ed). New York: Wiley.

Bonnecaze, R. T., Mano, N., Nam, B., & Heller, A. (2007). On the behavior of the porous rotating disk electrode. *Journal of The Electrochemical Society, 154*(2), F44-F47. http://dx.doi.org/10.1149/1.2403082

Cohen, Y., Avram, L., & Frish, L. (2005). Diffusion NMR spectroscopy in supramolecular and combinatorial chemistry: an old parameter—new insights. *Angewandte Chemie International Edition, 44*(4), 520-554. http://dx.doi.org/10.1002/anie.200300637

Conway, B. E. (1999). *Electrochemical Supercapacitors: Scientific Fundamentals and Technological Applications*. New-York: Kluwer-Plenum Press. http://dx.doi.org/10.1007/978-1-4757-3058-6

FreedomCAR Ultracapacitor Test Manual. (2004). Idaho National Engineering Laboratory Report. DOE/NE-ID-11173.

Hahn, M., Barbieri, O., Gallay, R., & Kötz, R. (2006). A dilatometric study of the voltage limitation of carbonaceous electrodes in aprotic EDLC type electrolytes by charge-induced strain. *Carbon, 44*(12), 2523-2533. http://dx.doi.org/10.1016/j.carbon.2006.05.002

Maletin, Y., Novak, P., Shembel, E., Izotov, V., Strizhakova, N., Mironova, A., ... & Podmogilny, S. (2006). Matching the nanoporous carbon electrodes and organic electrolytes in double layer capacitors. *Applied Physics A, 82*(4), 653-657. http://dx.doi.org/10.1007/s00339-005-3416-9

Maletin, Y., Podmogilny, S., Stryzhakova, N., Mironova, A., Danilin, V., & Maletin, A. (2008). Electrochemical double layer capacitor. *US Patent Appl No.* 20080151472.

Maletin, Y., Stryzhakova, N., Zelinsky, S., Gromadsky, D., Tychina, S., & Drobny, D. (2012). Can the best performance and improved design open the door to EDLC market? *Proc. 22nd Internat. Seminar on Double Layer Capacitor and Hybrid Energy Storage Devices*. Deerfield Beach, FL, 180-185.

Miller, J. R., Butler, S. M., Ryan, D. M., & McNeal, S. (2013). *Property, performance, and life of today's large-format electrochemical capacitors*. Proc. 3rd European Advanced Automotive Battery Conference, Strasbourg.   Retrieved   from http://us1.campaign-archive1.com/?u=84cc935cd75c22a368d1cd12e&id=31a3699821&e=193f657ac6

Price, W. S. (1997). Pulsed - field gradient nuclear magnetic resonance as a tool for studying translational diffusion: Part 1. Basic theory. *Concepts in magnetic resonance, 9*(5), 299-336. http://dx.doi.org/10.1002/(SICI)1099-0534(1997)9:5<299::AID-CMR2>3.0.CO;2-U

Strelko, V., Stryzhakova, N., Gozhenko, O., & Maletin, Y. (2009). N- and P-doped carbons as electrode materials. *Proc. 19th Internat. Seminar on Double Layer Capacitor and Hybrid Energy Storage Devices*. Deerfield Beach, FL, 115-123.

Tanner, J. E. (1970). Use of the stimulated echo in NMR diffusion studies. *The Journal of Chemical Physics, 52*, 2523. http://dx.doi.org/10.1063/1.1673336

14

15 and 16 Years-Old Students' Understanding of Factors That Influence Water Pollution

Neva Rebolj[1] & Iztok Devetak[1]

[1] University of Ljubljana, Faculty of Education, Kardeljeva pl. 16, 1000 Ljubljana, Slovenia

Correspondence: Iztok Devetak, University of Ljubljana, Faculty of Education, Kardeljeva pl. 16, 1000 Ljubljana, Slovenia. E-mail: iztok.devetak@pef.uni-lj.si

Abstract

Considering current environment condition, we should finally pay much more attention to environmental education. One of the methods, which would contribute to better environmental education, is of course including more environmental content to secondary school curriculum. Based on this, we would be able to expect the student's knowledge of water pollution factors to improve. 281 of first and second year students of health care school, nursing care program, participated in this research. Data were collected by analyzing test papers about water pollution factors. We have also made knowledge comparison about water pollution factors between 15 (first year of secondary school) and 16 year (second year of secondary school) old students. The results indicated that second year students achieved statistically significantly better results than first year students at the water pollution test paper. The difference results in the fact that second year students have already listened to the content of environmental pollution, which first year students had yet to hear. Results also show that first and second year students were more successful at answering general questions, which do not demand much of detailed environmental knowledge.

Keywords: environmental education, secondary school students' knowledge of water pollution factors

1. Introduction

1.1 Environmental Education

Environmental education (EE) is nowadays a topic of many seminars and conferences, where it is being discussed, how to bring knowledge of the environment and its laws closer to the people and that such education would have the effect of environmental protection of the individual, a society and a global population. In other words, environmental education seeks to empower individuals to understand environmental problems and to improve their ability to solve them (McCue, 2003). Environmental education has been suggested as one of the most effective ways to respond to environmental threats (Teksoz, Sahin, & Ertepinar, 2010). Many environmental education programs do not provide individuals with practical and useful information on how to deal with the environment in which they live. The Tbilisi Declaration, (www.naaee.org) and United Nations Environmental, Scientific, and Cultural Organization (www.naaee.org) agree the EE's goals should be focused on three fundamental aspects: (1) building awareness among individual citizens and community groups about the impact of the social, economic, political and ecological practices on the environment; (2) providing education opportunities for citizens so they acquire the necessary skills, knowledge, values and attitudes for environmental protection, and (3) fostering action-oriented behaviours towards environmental conservancy and sustainability. Consequently, the main emphasis of environmental education programs should be to change the individual's attitude towards the environment with a deeper understanding of the environment (O'Connor & Pooley, 2000). Environmental education is not limited only on formal education, but appears in many forms of informal education (Flowers, Guevara, & Whelan, 2009), for example: project work, workshops, field trips, museums, zoos, botanic gardens, etc. Similar views on the goals that environmental education should be accomplished in both formal and informal education have been submitted by others (E. Jeronen, J. Jeronen, & Raustia, 2008; Palmer & Neal, 1994). Most countries in the world understand the importance of EE, so it is incorporated into formal education. Most agree that EE is compatible with other science disciplines in the field of education, but there is much more emphasis on the central areas of science, such as biology, chemistry and physics. The first efforts to include environmental education in Latin America have been introduced in the form of informal educational programs

(González-Gaudiano, 2007). Currently in the United States, a lot of attention and support is devoted to include environmental education back into the curriculum (Campbell, Medina-Jerez, Erdogan, & Zhang, 2010). Davis and Elliott (2003) have found out that the field trip in kindergarten, as part of environmental education, particularly contributed to memorizing what has been heard and seen. After one year, children could still remember a lot of things from the field trip. Benton, Farmer and Knapp (2007) studied environmental education at an early age and made similar conclusions.

1.2 Teacher Education about Environment

In the world, Environmental Education is a priority in teacher education since the end of the twentieth century. New theories and techniques that have emerged in teacher education require the skills and knowledge used in practical terms (Teksoz et al., 2010). As stated by Miles and Cutter-Mackenzie (2006), teacher education for implementing environmental education in their classrooms can be defined as a "priority of priorities". Fien and Tilbury (1996) argue that the inclusion of environmental education in teacher education is "as an incentive for its introduction into the school curriculum" and in particular to "develop an effective course for training teachers in environmental education would result in innovation in the curriculum." Many researchers (Ballantyne, 1995; Cutter-Mackenzie, 2003; Jenkins, 1999-2000; Mastrilli, 2005; Mckeown-Ice, 2000; Powers, 2004; Spork, 1992; Tilbury, 1992, 1993, 1994) argue that the lack of pre-service and in-service teacher training in environmental education is the one that presents the major barrier that is preventing and/or limiting the effective implementation of environmental education in primary schools. From their research about the level of environmental knowledge of Michigan State University students, Kaplowitz and Levine (2005) concluded that increasing the level of environmental knowledge of tomorrow's teachers may be possible and also prosperous and that more research is needed to determine the role that teachers and their education play in the environmental education of their students. In the United States environmental education study, McKeown-Ice (2000) suggested, that there is a need for such studies to determine if correlations between environmental literacy, environmental education teaching competencies, and teacher preparation program level of involvement in environmental education exist. In the study about elements of success on environmental education, Theodore (2000) emphasized the importance of taking data from environmental education practitioners. He set three categories for developing elements of success in environmental education framework as: teaching conditions, teacher competencies and teaching practices. Environmental educators should possess the understandings, skills and attitudes associated with environmental literacy. These competencies have been defined in details in Excellence in Environmental Education—Guidelines for Learning (Pre K–12), published by the North American Association for Environmental Education (Teksoz et al., 2010). In three major teacher-training colleges, Goldman, Yavetz and Peer (2006) investigated the relationship between future teachers, environmental behaviour and background. They also worked on the environmental literacy of teacher training in Israel. Their study revealed that those future teachers manifest a low level of environmental literacy which was reflected in their environmental behaviour. Students majoring in fields related to the environment showed more knowledge about the environment compared to students of other disciplines. These findings are related to Arcury's (1990) views that "Increased knowledge about the environment is assumed to Change Environmental Attitudes" (p. 300). In Turkey, the Ministry of Education (MNE, 2005) only now plans to include environmental issues in science subjects in school curriculum. Some environmental issues such as endangered species, recycling, water pollution, energy use and deforestation are included in the newly developed curriculum of basic science. However, environmental education has been made a compulsory part of teacher education, despite the fact that new objects are linked to environmental issues. Therefore it became important to assess the environmental knowledge of pre-service teachers, and that environmental education should be made an integral part of teacher education (Teksoz et al., 2010).

2. Materials and Methods

2.1 Research Problem and Research Questions

In Slovenia, there is not much attention paid to the environmental education. In many high schools students do not learn much about the environment. We mean mostly secondary technical and vocational schools, where the emphasis is on the profession, so the majority of the lessons is devoted to practical training. In chemistry class the students should be taught the main causes of air, water and soil pollution and their consequences on people's health and lives. In healthy nutrition class they should be taught about the importance of clean drinking water intake and its laws (http://portal.mss.edus.si/msswww/programi2011/programi/Ssi/KZ-IK/katalog.htm). If environmental concepts are included in some high schools curriculum, there are usually not enough lessons to include environmental education into the actual classroom activities. On closer examination of biology and chemistry curriculum in vocational school, we can see that in the biology class students do not hear anything about the environment, while in the chemistry curriculum there is a section Chemistry and environment, where students learn some basic concepts about the contaminants in air, soil and water (Poberžnik et al., 2007; Zupančič, Vičar, Gobec, & Mršić, 2007). Teachers also play an important role in educating students by setting an example of how to deal with environmental issues. Two basic research questions are based on the research problem described above: (1) How deeply do students aged 15 to 16 years understand the factors that contribute to water pollution? (2) Are the differences in the knowledge of first and second year students statistically significantly different?

2.2 Sample

The study involved 281 students, of which 144 (51%) were first-year students and 137 (49%) were second year students of secondary medical school, majoring in Nursing. The sample comprises 2/3 female and 1/3 male students. Students came from different parts of Slovenia. They first encountered chemistry concepts (also connected with environmental issues) in the first and second periods of the primary school (age 6 to 11) and more in-depth in 105 hours of Science in 7th grade (age 12) and 70 hours of chemistry in eighth grade and 64 hours of chemistry in ninth grade (age 13 and 14). In elementary school students also had a chance to select an optional subject such as Experiments in chemistry, Chemistry in life or Chemistry in the environment, in the extent of 35 hours and also Environmental Education I, II and III. In a secondary medical school students only had 70 hours of chemistry in the first year. Similarly students first came across biology in first and second period of elementary school and studied it more thoroughly in the third period (grades 7 to 9). They had 52.5 hours of biology in eighth grade and 70 hours in ninth grade. They could also choose an optional subject Organisms in natural and artificial environment in the extent of 35 hours in seventh and eighth grade and 32 hours in ninth grade. In secondary medical school students only had 70 hours of biology in the first year.

2.3 Instrument

Data were collected with a quantitative data collection technique; students answered the test paper questions. The test comprised seven multiple-choice questions with the possibility of their opinion, which was not in the answers. Overall, students could achieve 7 points. The questionnaire covered some well known issues and also some issues where it was necessary to show some knowledge. As well known issues we asked students what they think is polluting the river the most and what can be inferred if there are no fish in the water. The questions that required some knowledge were the following; we were interested in which substances are present in the water, if the analysis shows the presence of ammonia and nitrate. What does it mean that the river has a self-cleaning function and what happens if there is an excessive growth of green algae in the river. We were also interested in what parameters can be found with physical indicators of water pollution.

2.4 Research Design

In this study we used the descriptive method. Test was applied in a group during the class under standard conditions. They needed an average of 10 minutes to finish it. The test was anonymous and the data was used only for the purposes of this study. The data were statistically analyzed using computer program Excel and the statistical program Statistical Package for the Social Sciences (SPSS). The following tests were used for the analysis of the results within the program SPSS: t-test, χ^2-test (this test of independence was used because we have two groups of students and five possible values of nominal variables) and frequencies of responses in each item.

3. Results and Discussion

The results show that there is a statistically significant difference in students' understanding of water pollution factors between the first- (mean (M) = 3.8, standard deviation (SD) = 1.3) and second-year students (M = 4.14, SD = 1) (t (279) = - 2.142, p = 0.03). On average, first-year students reached 54.3% of total points and second-year students 59% of total points. The students' achievements show quite average knowledge of factors that cause the water pollution. Second-year students understand the water pollution factors a bit better, as they had chemistry in the first year of secondary school, where they had conducted experiments on water pollution and have acquired knowledge by practical work, which they have memorized. By the time the research was conducted, first-year students have not yet been involved into the contents of chemistry and biology, which would relate only to water pollution and its consequences on the organisms. The reason for a weaker performance of the first-year students could also be that they have forgotten certain information related to the river pollution from the primary school, but the second-year students' achievements were not much better than their first-year peers. Table 1 shows the detailed analysis of the differences between students' achievements in the first- and second-year.

Table 1. Proportion of the first- and second-year students' achievements and the significance of the difference in achievements between them at the specific item

Item[a]	Grade 1		Grade 2		The differences between 1st and 2nd grade students' achievements	$\chi^{2\,b}$	p
	f	%	f	%			
Item 1	55	19.6	70	24.9	+5.3	4.731	**0.03**
Item 2	88	31.3	103	36.7	+5.4	6.385	**0.01**
Item 3	31	11.0	25	8.90	−2.1	0.473	0.49
Item 4	88	31.3	82	29.2	−2.1	0.046	0.83
Item 5	130	46.3	128	45.6	−0.7	0.929	0.34
Item 6	63	22.4	79	28.1	+5.7	5.438	**0.02**
Item 7	90	32.0	78	27.8	−4.2	0.905	0.34

Notes:

[a] Items:

Item 1: What do you think is polluting our water the most?

Item 2: What could be in the water, if the analysis of the water proves the presence of ammonia?

Item 3: What could be in the water, if the analysis of the water proves the presence of nitrates?

Item 4: What happens to the animals in the river when the excessive growth of green algae appears?

Item 5: What can you conclude if there are no fish in the water?

Item 6: What does it mean that the river has a self-cleaning function?

Item 7: What can be found with chemical indicators of water pollution?

[b] $df = 1$ at all χ^2 tests.

Statistically significant differences ($p < 0.05$) in the understanding of factors that influence water pollution, depending on the students' year of schooling, have been identified in questions 1, 2 and 6. Second-year students answered all three questions better than first-year students. From these results we can conclude that second-year students are more familiar with the water pollution factors. In general, the first-year students were better at answering the questions, since the proportion of the correct answers was better in four out of the seven questions, but these differences are statistically not significant. In both grades, students achieved the best result in item 5, which relates to what may be inferred if we do not see any fish or plants in the water. The worst result was also achieved by both grades in item 3, which was asking about knowing the importance of nitrates in the water. Analysis of the data showed that students in both grades answered more accurately at more general questions, whose content is often presented in the form of informal education, such as workshops, project work, field trips … Items that demanded more specific knowledge, especially those including chemistry were the ones where the answers were incorrect in more cases. Table 2 shows the frequency and ratio of different answers to multiple choice items for each grade.

Table 2. Frequencies and proportions of individual responses in both grades to the first item

What do you think is polluting our water the most?	Households		Municipal sewage		Industry		Agriculture	
	f	%	f	%	f	%	f	%
Grade 1	15	10.0	69	48.0	55.0	38.0	5	3.5
Grade 2	14	10.2	50	36.5	70.0	51.1	3	2.2

In the first item students had to decide about who or what is polluting water the most. First-year students most frequently selected municipal sewage, while second-year students mostly selected industry. Municipal sewage and the industry are those two sources of water contaminants, which are most frequently exposed and are frequently present in the media. From the results we can conclude that students monitor the media and are aware of the most common factors of water pollution.

Table 3. Frequencies and proportions of individual responses to the second item in both grades

What could be in the water, if the analysis of the water proves the presence of ammonia?	Slurry		Too many plants		Too many animals		Washing powder	
	f	%	f	%	f	%	f	%
Grade 1	88	61.1	12	8.3	3	2.1	41	28.5
Grade 2	103	75.2	11	8.0	5	3.6	18	13.1

The second item (Table 3) was asking the students about what is the most possible source of ammonia in the water, if the analysis proves the presence of ammonia. Students of both grades have most often selected the answer slurry. It can be concluded that the students are familiar with the problem of negligent use of manure on farmland and other land used for agriculture and that slurry is a possible source of ammonia in the water, not only fertilizers that are used in the extensive agriculture.

Table 4. Frequencies and proportions of individual responses to the third item in both grades

What could be in the water, if the analysis of the water proves the presence of nitrates?	Fertilizers		Salts		Acids		Plants	
	f	%	f	%	f	%	f	%
Grade 1	31	21.5	64	44.4	49	34.0	0	0.0
Grade 2	25	18.2	70	51.1	37	27.0	5	3.6

The third item (Table 4) was asking about what is the most possible source of nitrates in the water, if the analysis proves the presence of nitrates. Students of both grades mostly selected the answer salts, which is obviously not correct. If the water analysis shoves the presence of nitrates, it is a result of fertilizers in the water. The conclusion is that students are not familiar with the problem of excessive use of fertilizers in agriculture, and that they connect the nitrates with salts. Chemically nitrates are salts, but the sources of nitrates in the water are not salts in general, but fertilizers.

Table 5. Frequencies and proportions of individual responses to the fourth item in both grades

What happens to the animals in the river when the excessive growth of green algae appears?	Animals in the water may not reproduce, because they do not have the space.		Animals in the water die because the decomposition of the plants consumes oxygen.		The number of animals in the water is increased because they have a lot of food.		The animals are moving to another river.	
	f	%	f	%	f	%	f	%
Grade 1	20	13.9	88	61.1	22	15.3	14	9.7
Grade 2	18	13.1	82	59.9	30	21.9	7	5.1

The fourth item (Table 5) was related to the fate of the animals in the river if the excessive plant growth appears because of the high concentration of fertilizers and plants' decomposition after dying. Also here the students of both grades share the same opinion, since the most common answer was that animals die in water because the decomposing plants consume oxygen. It is encouraging that students understand the meaning of the eutrofication of the surface waters and that decomposition of the plants with bacteria causes the decrease of dissolved oxygen in the water. Because of this the animals can suffocate.

Table 6. Frequencies and proportions of individual responses to the fifth item in both grades

What can you conclude if there are no fish in the water?	The water is too polluted for fish and plants.		The water is too cold for fish and plants.		The water is rushing too much for fish and plants.		The water is too warm for fish and plants.	
	f	%	f	%	f	%	f	%
Grade 1	130	90.3	7	4.9	6	4.2	1	0.7
Grade 2	128	93.4	4	2.9	2	1.5	3	2.2

The fifth item (Table 6) was asking the students about what they can conclude if there are no fish and plants in the water. Most of the students from both groups selected the same answer, the water is too polluted for fish and plants. This result indicates, that students can connect lifeless ecosystem with pollution. The discussion in media about the pollution is usually connected with the negative effects on the living organisms in the water. Media also report massive deaths at least two times a year in Slovenia, so students in secondary school undoubtedly came across this information at least in the media reports if not in the formal education.

Table 7. Frequencies and proportions of individual responses to the sixth item in both grades

What does it mean that the river has a self-cleaning function?	The rivers clean themselves by disposing of the waste materials on the banks.		The rivers clean themselves and neutralize the waste material by microorganisms that live in them.		The rivers carry away waste materials into the sea.		The rivers dispose of the waste materials on the bottom of riverbed.	
	f	%	f	%	f	%	f	%
Grade 1	57	39.6	63	43.8	10	6.9	14	9.7
Grade 2	30	21.9	79	57.7	13	9.5	15	10.9

In the sixth item (Table 7) students had to select the right answer about the rivers' self-cleaning abilities. Students from the first and second grade most frequently selected the answer B, which explained that the rivers themselves degrade and neutralize the waste material by microorganisms that live in the water and on the bottom of the rivers. In the first grade less than 50% of students understand the rivers' self-cleaning abilities, but in the second grade of the secondary school more students (almost 60%) know that rivers can clean themselves. But on the other hand, quite a lot of students, almost 40% in the first and 22% in the second grade think that rivers clean themselves by disposing the waist material on its banks. This is also understandable, because students can see different material, especially plastics that are disposed by the high waters, on the river banks.

Table 8. Frequencies and proportions of individual responses to the seventh item in both grades

What can we find with chemical indicators of water pollution?	The presence of microorganisms.		The presence of plants and animals.		Colour, odour and clarity of the water.		pH, hardness, presence of different substances in water.	
	f	%	f	%	f	%	f	%
Grade 1	21	14.6	9	6.3	25	17.4	89	61.8
Grade 2	18	13.1	5	3.6	36	26.3	78	56.9

The seventh question was asking the students if they know what we can find with chemical indicators of water pollution. Students of both grades have tended to choose the answer d, pH, water hardness, the presence of different substances in water. It is interesting that first grade students were more successful at the last item, than their second grade counterparts. It is true that the difference is not significant, but students in the first grade identified the analysis of the different substances in the water as the chemical analysis. This is in consistence with the fact that first grade students were involved in the chemistry education in the time of the research, but one year has passed since the second year students participated at the chemistry lessons.

4. Conclusions

The curriculum of a secondary medical school contains environmental content only as a part of the chemistry, but in a very small extent (e.g. know the main air pollutants; carbon dioxide, sulphur dioxide, nitrogen oxides, ozone, chlorofluorocarbons (CFCs), smog and the consequences of air pollution; acid rain, greenhouse effect, ozone layer destruction, know the drinking water main pollutants; phosphates, nitrates, pesticides, exploring the resources and consequences of major pollutants of soil; fertilizers, biocides, detergents, petroleum products ...) (http://portal.mss.edus.si/msswww/programi2011/programi/Ssi/KZ-IK/katalog.htm). Most often the content is related to general knowledge on environmental issues. General knowledge of the factors of river pollution can be assessed as good. We expected that the knowledge of the factors in our education had improved, which we can prove and confirm with our research, which showed that the students of the second grade were more successful than the students of the first grade, because they have already had biology and chemistry classes. We found out, that there are statistically significant differences between the knowledge of the river pollution of the first-grade students and the second-grade students. But on the other hand it should be emphasised that the average knowledge of the topic is still quite average and it should be improved. It is expected that the knowledge of river pollution factors is different between boys and girls. It would be advisable to carry out a research that would evaluate the difference in knowledge according to gender and the factors that influence the gender differences. Researches (González-Gaudiano, 2007; Campbell et al., 2010) show that environmental contents are interesting for the students` population. In Slovenia these topics are insufficiently integrated into the chemistry, biology, nutrition and ultimately health care curriculums. Environmental context has a great effect on healthy lifestyle, health and general welfare of the individuals. It would be sensible to add a subject to the curriculum of secondary medical schools that would acquaint students with the most pressing problems of pollution of our environment which is also one of the key factors that greatly affect the health and welfare of the individual. The problem is that there are constantly new topics that could be integrated into the curriculum, but in the end only few are (Yueh, Cowie, Barker, & Jones, 2010). Because of this teachers of the specific subject should decide which topics are the most interesting and have the potential to motivate students to learn science. In an effort to promote the emergence of a subject, such as environmental education, it is necessary to consider the participation of the local environment, links with a national network of cooperation and, ultimately, participation of a school. Without these factors, the fight for integration of the subject in the curriculum is difficult (Yueh et al., 2010). Conde and Sánchez (2010) describe the various ways in which environmental education should be included in the curriculum. One way is the so-called "sword" model, which touches on all curriculum areas as a supplement in the form of isolated or occasional workshops, activities, either inside or outside the school context. Pujol and Bonil (2003) see a "greening" of the curriculum as a complex and dynamic process, built on three main foundations: (1) a new collective ethic, (2) a new style of thinking and (3) a new form of transformation of operation. It may be worthwhile to conduct a research among chemistry and biology teachers, as this would determine how they themselves deal with the environmental issues and how do they integrate these topics into their teaching. On this basis, we would get a clearer picture of why some students know the factors of pollution much better than others and what are the factors that influence the students' understanding of the environmental issues. It would be also important to develop a model for applying the environmental content into the science subject in Slovenian schools. This model could be developed according to the extensive results obtained by the study of the wide range of different variables that influence teaching and learning of environmental issues.

References

Arcury, T. (1990). Environmental attitude and environmental knowledge. *Human Organization, 49*(4), 300-304.

Ballantyne, R. (1995). Environmental teacher education: Constraints, approaches and course design. *International Journal of Environmental Education and Information, 14*(2), 115-128.

Benton, G. M., Farmer, J., & Knapp, D. (2007). An Elementary School Environmental Education Field Trip: Long-Term Effects on Ecological and Environmental Knowledge and Attitude Development. *Journal of Environmental Education, 38*(3), 33-42. http://dx.doi.org/10.3200/JOEE.38.3.33-42

Campbell, T, Medina-Jerez, W., Erdogan, I., & Zhang, D. (2010). Environmentally educated teachers: The priority of priorities? Connect,XV(1), 1-3. Exploring science teachers' attitudes and knowledge about environmental education in three international teaching communities. *International Journal of Environmental & Science Education, 5*(1), 3-29.

Conde, M., & Sánchez, J. S. (2010). The school curriculum and environmental education: A school environmental audit experience. *International Journal of Environmental & Science Education, 5*(4), 477-494.

Cutter-Mackenzie, A. (2003). *Eco-Literacy: The "Missing Paradigm" in Environmental Education.* Unpublished doctoral dissertation, Central Queensland University, Brisbane.

Davis, J. M., & Elliott, S. (2003). Early childhood environmental education: Making it mainstream. Research into Practice. *Early Childhood Australia, Inc., the Journal of Environmental Education, 38*(3), 33-42.

Fien, J., & Tilbury, D. (1996). *Learning for a sustainable environment: An agenda for teacher education in Asia and the Pacific.* Bangkok: UNESCO.

Flowers, R., Guevara, J., & Whelan, J. (2009). Popular and Informal environmental education – The need for more research in an 'emerging' field of practice Report. *Zeitschrift fur Weiterbildungsforschung, 32*(2), 36-50.

Goldman, D., Yavetz, B., & Peer, S. (2006). Environmental literacy in teacher training in Israel: Environmental behavior of new students. *Journal of Environmental Education, 38*(1), 3-22. http://dx.doi.org/10.3200/JOEE.38.1.3-22

González-Gaudiano, E. (2007). Schooling and environment in Latin America in the third millennium. *Environmental Education Research, 13*(2), 155-169. http://dx.doi.org/10.1080/13504620701295684

Jenkins, K. (1999-2000). Listening to secondary pre-service teachers: Implications for teacher education. *Australian Journal of Environmental Education, 15*(16), 45-56.

Jeronen, E., Jeronen, J., & Raustia, H. (2009). Environmental education in Finland - A case study of environmental education in nature schools. *International Journal of Environmental and Science Education, 4*(1), 1-23.

Kaplowitz, M., & Levine, R. (2005). How environmental knowledge measures up at a Big Ten university. *Environmental Education Research, 11*(2), 143-160. http://dx.doi.org/10.1080/1350462042000338324

Mastrilli, T. (2005). Environmental education in Pennsylvania's elementary teacher education programs: A statewide report. *Journal of Environmental Education, 36*(3), 22-30. http://dx.doi.org/10.3200/JOEE.36.3.22-30

McCue, C. (2003). Environmental Education. *Environmental Encyclopedia.*

McKeown-Ice, R. (2000). Environmental education in the United States: A survey of pre-service teacher education programs. The *Journal of Environmental Education, 32*(1), 4-12. http://dx.doi.org/10.1080/00958960009598666

Miles, R., & Cutter-Mackenzie, A. (2006). Environmental Education: Is it really a Priority in Teacher Education? In S. Wooltorton, & D. Marinova (Eds.), *Sharing wisdom for our future, Environmental education in action.* Proceedings of the National Conference of the Australian Association for Environmental Education. Sydney: Australian Association for Environmental Education (1-6).

Ministry of National Education of Turkey [MNE]. (2005). *İlköğretim fen ve teknoloji dersi (6-8 ınıflar) öğretim programı (Elementary school science and technology curriculum (grades 6-8).* Ankara, Turkey.

North American Association of Environmental Education [NAAEE]. (2009). The Tbilisi Declaration: Final report intergovernmental conference on environmental education. Tbilisi, USSR, 14-26. October 1977, Paris, France. Retrieved March 10, 2009, from http://www.naaee.org

O'Connor, M., & Pooley, J. A., (2000). Environmental Education and Attitudes: Emotions and Beliefs are What is Needed. *School of Psychology at Edith Cowan University in Perth, Australia, 32*(5), 711-723.

Palmer, J., & Neal, P. (1994). *The handbook of environmental education.* London: Routledge.

Poberžnik, A., Turk, M., Malek, N., Kožlakar, R., Banik, A., & Skvar, M. (2007). Katalog znanja kemija. [Chemistry knowledge catalogue]. Retrieved February 9, 2011, from http://www.zrss.si

Powers, A. L. (2004). Teacher preparation for environmental education: Faculty perspectives on the infusion of environmental education into pre-service methods courses. *The Journal of Environmental Education, 35*(3), 3-12.

Pujol, R. M., & Bonil, J. (2003). Una propuesta de ambientalización curricular desde la formación científica: el caso del crecimiento urbano. En M. Junyent, A. Mª. Geli y E. Arbat (eds.), *Ambientalización Curricular de los Estudios Superiores. 2: Proceso de Caracterización de la Ambientalización Curricular de los Estudios Universitarios* (pp. 151-171). Girona: Universidad de Girona - Red ACES.

Spork, H. (1992). Environmental education: A mismatch between theory and practice. *Australian Journal of Environmental Education, 8,* 147-166.

Teksoz, G., Sahin, E., & Ertepinar, H. (2010). A new vision for chemistry education students: Environmental education. *International Journal of Environmental & Science Education, 5*(2), 131-149.

Theodore, S. M. (2000). Elements of success in environmental education through practitioner eyes. *Journal of Environmental Education, 31*(3), 4-11. http://dx.doi.org/10.1080/00958960009598639

Tilbury, D. (1992). Environmental education within pre-service teacher education: The priority of priorities. *International Journal of Environmental Education and Information, 11*(4), 267-280.

Tilbury, D. (1993). A grounded theory of curriculum development and change in environmental education at the teacher education level. Unpublished paper presented at the UNESCO Asia – Pacific Region Seminar on Environmental Education and Teacher Education, Griffith University.

Tilbury, D. (1994). The International development of environmental education: A basis for a teacher education model? *The International Journal of Environmental Education and Information, 13*(1), 1-20.

Yueh, M. M., Cowie, B., Barker, M., & Jones, A. (2010). What influences the emergence of a new subject in schools? The case of environmental education. *International Journal of Environmental & Science Education, 5*(3), 265-285.

Zupančič, G., Vičar, M., Gobec, K., & Mršić, H. (2007). Katalog znanja biologija [Biology knowledge catalogue]. Retrieved February 9, 2011, from http://www.zrss.si

Nutrient Removal From Waste Water by Macrophytes – An Eco-Friendly Approach to Waste Water Treatment and Management

Sukhen Roy[1], J. K. Biswas[1] & Sanjay Kumar[2,3]

[1] Department of Environmental Management, University of Kalyani, West Bengal, India

[2] Department of Physics, B. R. Ambedkar Bihar University, Muzaffarpur, Bihar, India

[3] Centre for Renewable Energy and Environmental Research, Muzaffarpur, Bihar, India

Correspondence: Sanjay Kumar, Centre for Renewable Energy and Environmental Research, Kafen Cottage, Balughat, Muzaffarpur, Bihar, India. E-mail: prof.kumars@gmail.com

Abstract

Wastes are resources out of place. Waste management technique essentially includes nutrient removal for better uses through low cost sustainable eco-technology. In this paper, local indigenous Macrophytes (*Ipomoea aquatic*, *Trapa*, *Nymphaearubra*, and *Pistia* sp.) are investigated to assess their ability to reclaim nutrients and remove pollutants from sewage water in West Bengal. Nitrogen, phosphorus, BOD and salinity characteristics are investigated. These Macrophytes have adopted the environment and are found to grow in the vicinity of wastewater command areas. Results indicate that their potential is very high in removing nutrient pollutants cost-effectively. Nutrient thus removed can be used in cultivation process, reducing demand of chemical fertilizers.

Keywords: macrophytes, waste water management, pollutant, nutrient removal

1. Introduction

Urban India generates over 20 billion litres of sewage. Almost all of them ultimately find their way into aquatic ecosystem, seriously damaging them. Domestic wastes even pose eco-toxicological risk and health hazards when it unknowingly or accidently intermix with industrial effluents. At the same time, wastes are also resources out of place. It should not remain unutilized, but be returned to earth system for various human welfare activities through reclamation mechanisms and appropriate management strategies. About twenty different techniques are being applied all over the world (Welch, 1996; Debusk, Reddy, & Clough, 1989) for waste water treatment. These include diversion of nutrient inputs and its treatment, inactivation of nutrients in input water, hypolimnetic aeration and artificial bubbling, chemical controlling and bio-manipulation approaches.

Nutrient diversion through constructed wetlands have become an important tool in developed countries for efficient management of wastewater during the last decade (Brix, 1994; Kadlec & Knight, 1994; Knight, Kadlec, & Reed, 1992). However, economic feasibility of such systems limits their application, especially in developing countries (Sun & Qu, 1998). Traditional wastewater treatment plants are also not very cost-effective as they consume large amount of resources for construction and maintenance. Besides, they use hazardous chemicals and produce contaminated or even toxic sludge as a byproduct, thus are counterproductive in several cases (Etnier & Guterstam, 1996).

During last decade, there have been major changes in the reclamation processes of wastewater, employing environment friendly low cost sustainable techniques such as integration of Macrophytes (Mitsch, 1996; Mohanty & Sinha, 1999). In the aquatic environment, macrophytes are known for their capacity to purify waste water (Gumbricht, 1993; Ozimek, Van Donk, & Gulati, 1993). Besides, they enlarge the matrix for many bacteria and zoobenthos and improve the condition for their living and reproduction. They also provide a sub-system to purify wastewater and promote the cycling of substances in water body (Busnardo, Gersberg, Langis, Sinicrope, & Zedler, 1992; Poole, 1996; Yan & Ma, 1998). Macrophytes base treatment system can be divided into free-floating ponds,submerged and constructed wetlands with emergent macrophytes depending on their structure and function. It can be classified in two groups, (a) Free floating and submerged species and, (b) constructed wetlan. The later is divided into subsystems - free water surface system and subsurface flow system.

However, all these systems are primarily based on either monoculture or polyculture of vascular plants in shallow eutrophic water bodies which receive wastewater with a long residence time relative to that of conventional wastewater treatment systems.

In this paper, Macrophytes, *Ipomoea aquatic*, *Trapa*, *Nymphaea rubra*, and *Pistia* sp. are selected and investigated to assess their ability to reclaim nutrients (nitrogen and Phosphorus) and remove pollutants (BOD) from sewage water in West Bengal. These Macrophytes are indigenous, have adopted the environment and are found to grow in the vicinity of wastewater command areas.

2. Materials and Method

2.1 Site Description

Kalyani Sewage Treatment Plant caters to a population of 200,000 at Kalyani, West Bengal (22°58'30"N, 88°26'04"E, elevation – 11 meters). Treatment ponds associated with Kalyani Sewage Treatment Plant are Anaerobic type and Facultative type. Domestic wastewater discharge contained both organic and inorganic solids in dissolved and suspended forms and is relatively dilute with 85–95% water. Preliminary investigation indicated 250–400 ppm of organic C and 80–120 ppm of total nitrogen, thus giving a C:N ratio of around 1:3. The composition is qualitatively different from industrial wastewater in which heavy metals and toxic chemicals dominate.

2.2 Macrophytes Selection

Following criteria is used in selecting macrophytes,

- adaptability to local climate;
- tolerance to adverse climatic conditions;
- tolerance to adverse concentration of pollutants;
- pollutants assimilative capacity;
- high rate of photosynthesis;
- high oxygen transport capability;
- resistance to pests/diseases;
- ease of management harvesting.

Based on these criteria, the macrophytes selected in the study are *Ipomoea aquatic* (Figure 1), *Trapa* (Figure 2), *Nymphaea rubra* (Figure 3), and *Pistia* sp. (Figure 4). Their observed growth in sewage disposal area is remarkable. Other properties are as follows;

Water Spianch (*Ipomoea aquatica*) – It is a semi-aquatic, tropical plant grown as a leafy vegetable in water or on moist soil. Its stems are 2–3 metres long and hollow, rooting is at the nodes. Leaf shape and size vary from typically sagittate (arrow head-shaped) to lanceolate, 5–15 cm long and 2–8 cm broad. The flowers are trumpet-shaped, 3–5 cm diameter, usually white in colour with a mauve centre. The flowers can form seed pods.

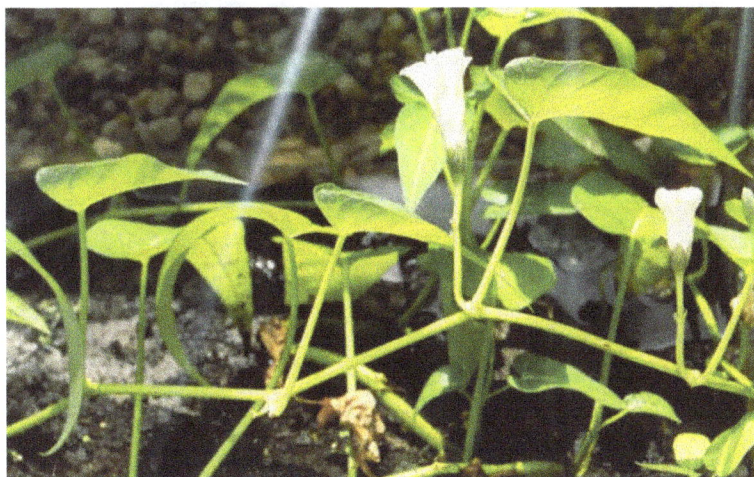

Figure 1. Water Spianch (*Ipomoea aquatica*)

Trapa – Two species of trapa is found in Kolkata region: *Trapa natans and Trapa bicornis*. Both the species are native to warm temperate parts of Eurasia and Africa. It is a floating annual aquatic plant. It grows in slow-moving water. It can survive in 5 meters deep water body. They bear ornately shaped fruit containing a single very large starchy seed.

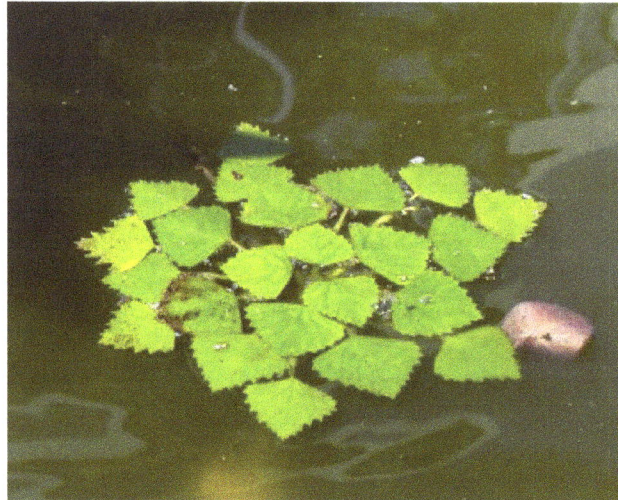

Figure 2. Trapa (*Trapa natans and Trapa bicornis*)

Pistia (*Pistia stratiotes*) – It is a perennial monocotyledon with thick, soft leaves that form a rosette. It floats on the surface of the water and its hanging roots remain submersed beneath floating leaves. Leaves (~ 14 cm long) have parallel veins and wavy margins, covered in short hairs forming basket-like structures, which trap air bubbles, increasing the plant's buoyancy. Flowers are dioecious. It can undergo asexual reproduction.

Figure 3. *Pistia* (*Pistia stratiotes*)

Water Lily (*Nymphaea Rubra*), family *Nymphaeaceae*: Its leaves have a radial notch from the circumference to the petiole in the center. It is closely related to Nuphar, another genus commonly called "lotus". However, its flower petals are much larger than the sepals and fruits are held above water level to maturity; in contrast with smaller sepals of Nuphar and underwater fruits.

Figure 4. Water Lily (*Nymphaea Rubra*)

2.3 Experimental Set up

Sixteen sets of water container of capacity 20 lts and height 40 cm were prepared. 10 containers were used for facultative sample treatment and six for anaerobic sample treatment. 6 cm of the bucket were filled with mud from respective ponds to provide actual condition of growing macrophytes. Sewage water from facultative and anaerobic pond was collected. Containers with facultative and anaerobic muds were filled with 12.5 lts of respective sewage water collected. Two sets of container with facultative pond samples were planted with *Ipomoea aquatic*, *Trapa* (*Trapa natans* and *Trapa bicornis* mixed), *Pistia stratiotes*, *Nymphaea rubra*, and similarly, two sets of container with anaerobic pond samples were planted with *Ipomoea aquatic*, *and*, *Pistia stratiotes*. Two sets of container with facultative and anaerobic sewage samples were used as indicators to discount possibility of other mechanisms such as photo-degradation etc. Table 1 shows the list of macrophytes used in various containers and their symbols. In anaerobic condition only two macrophytes showed potential of nutrient removal in initial experiments. These two species possibly increased aeration and absorption of nutrients. Evaporation of water is compensated with distilled water supplement. The study was carried out for thirty days in May-June 2013, and samples were collected on Day 0, 1, 3, 5, 7, 10, 15, 20 and Day 30. These samples were instantly analysed to collect data without any loss.

Nitrogen and Phosphorous contents were measured by UV Spectrophotometer, extracting very small amount of sample from the container. pH and salinity were measured with Eco Tester meter and PCS Tester 3. BOD was measured with WTW OXITOP IS-6 BOD bottles with digital display. Appropriate reagents for BOD, Nitrate, Nitrite, Ammonical Nitrogen and phosphate determination were used.

Table 1. List of macrophytes and respective ponds

Symbols	Pond Type	Macrophyte used
AF	Facultative	*ipomoea aquatica*
BF	Facultative	*trapa natans*
CF	Facultative	NONE
DF	Facultative	*nymphae rubra*
EF	Facultative	*Pistia* sp.
AA	Anaerobic	*ipomoea aquatica*
CA	Anaerobic	NONE
EA	Anaerobic	*Pistia* sp.

3. Results and Discussion

Figures 5 and 6, shows the variation of nitrite concentration in the containers. There is marginal decrease in nitrite concentration in the indicator container due to unknown factors. Even after discounting this marginal decrease, macrophytes are very effective in absorption of nitrites. In facultative pond, *Ipomoea aquatic* and *Pistia stratiotes* shows very high potential. 15 days is good enough to absorb over 65% of nitrite after discounting for other factors responsible for reduction in nitrite concentration. These two species are even more active in anaerobic ponds. Nitrite absorbed by the plant can be gainfully used as natural fertilizer.

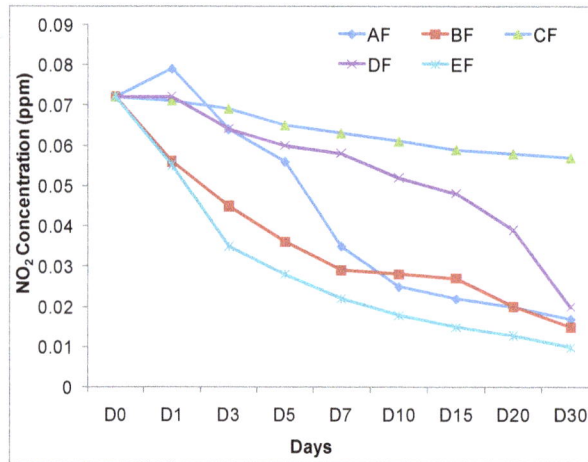

Figure 5. Variation of NO_2 concentration (ppm) in facultative pond

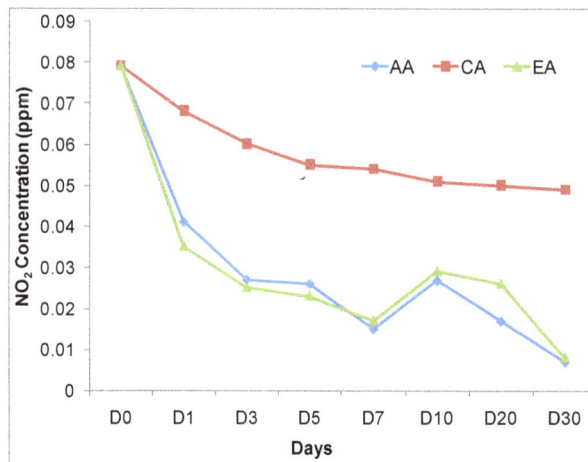

Figure 6. Variation of NO_2 concentration (ppm) in anaerobic pond

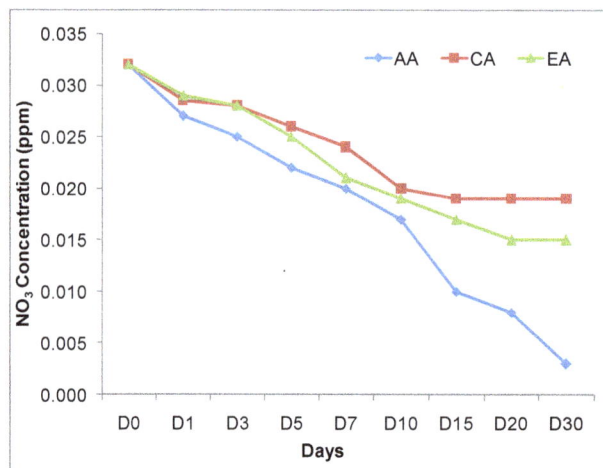

Figure 7. Variation of NO_3 concentration (ppm) in facultative pond

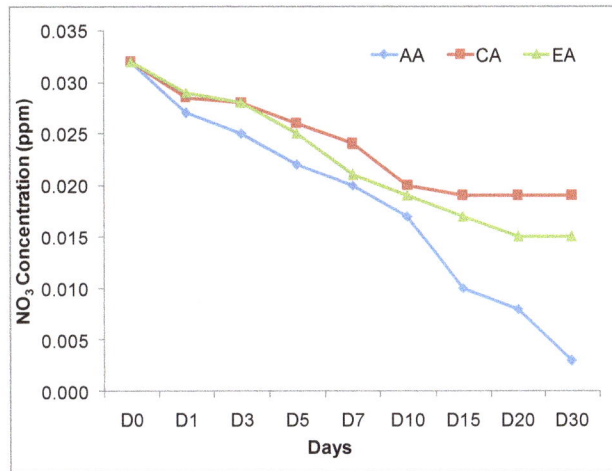

Figurer 8. Variation of NO$_3$ concentration (ppm) in anaerobic pond

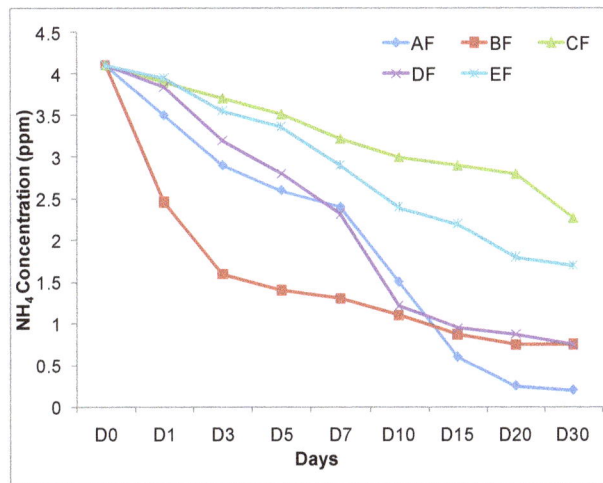

Figure 9. Variation of NH$_4$ concentration (ppm) in facultative pond

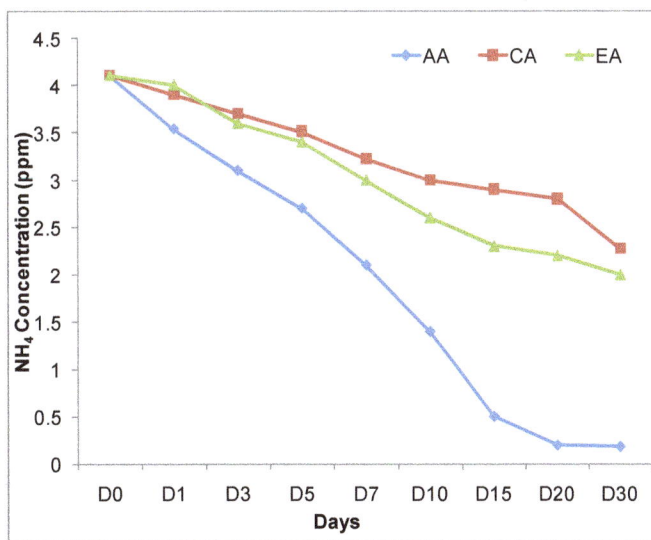

Figure 10. Variation of NH4 concentration (ppm) in anaerobic pond

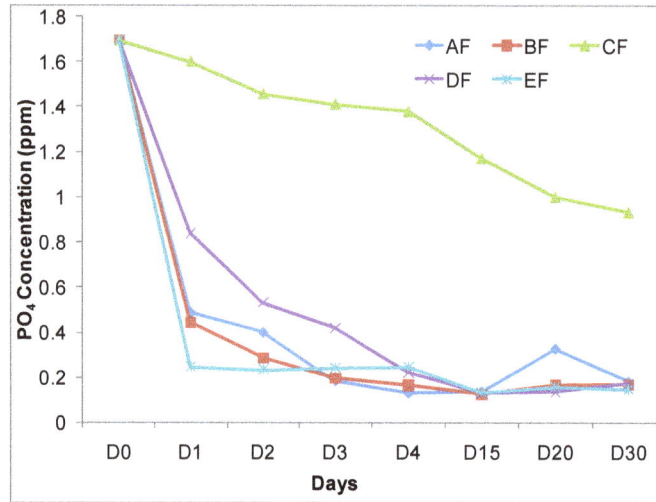

Figure 11. Variation of PO$_4$ concentration (ppm) in facultative pond

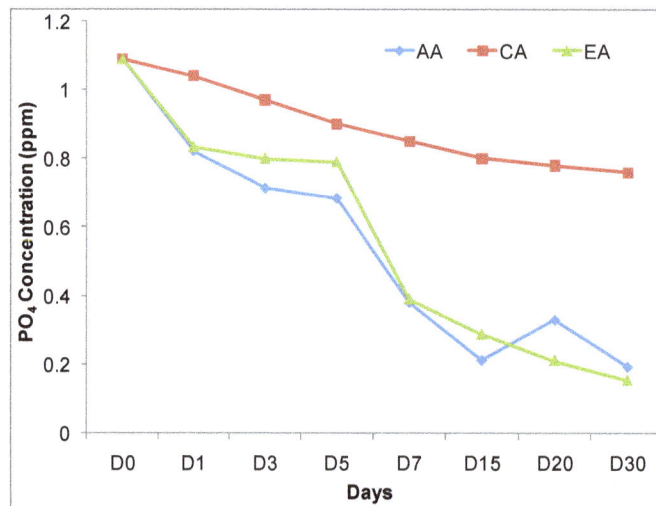

Figure 12. Variation of PO$_4$ concentration (ppm) in anaerobic pond

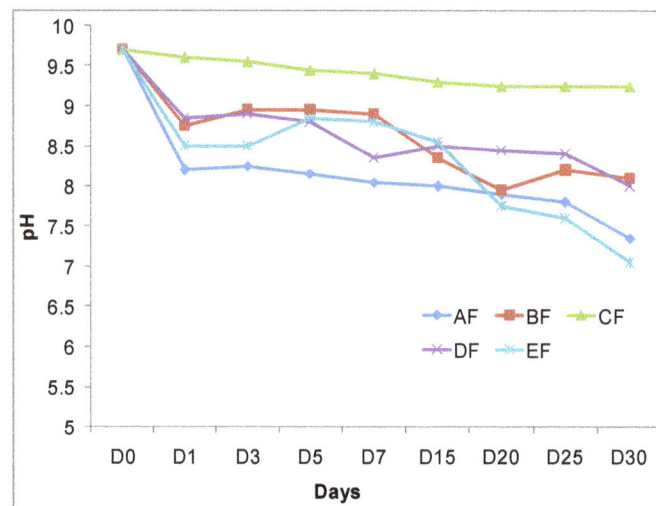

Figure 13. Variation of pH concentration in facultative pond

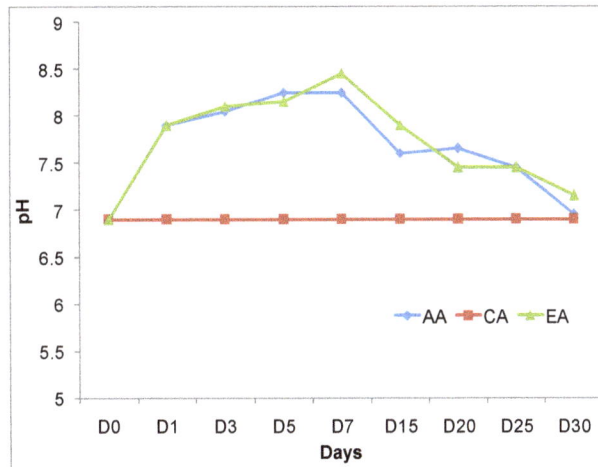

Figure 14. Variation of pH concentration in anaerobic pond

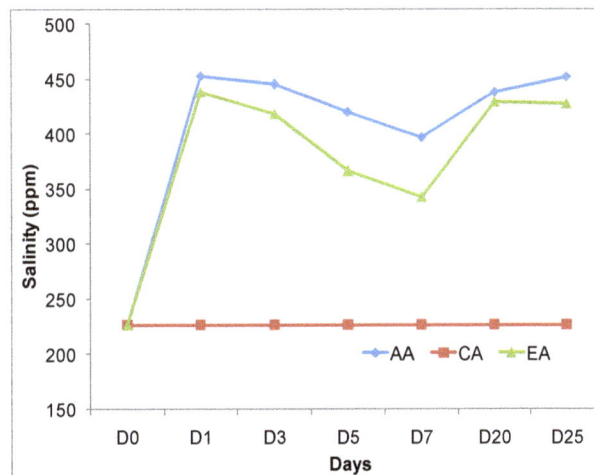

Figure 15. Variation of Salinity (ppm) in anaerobic pond

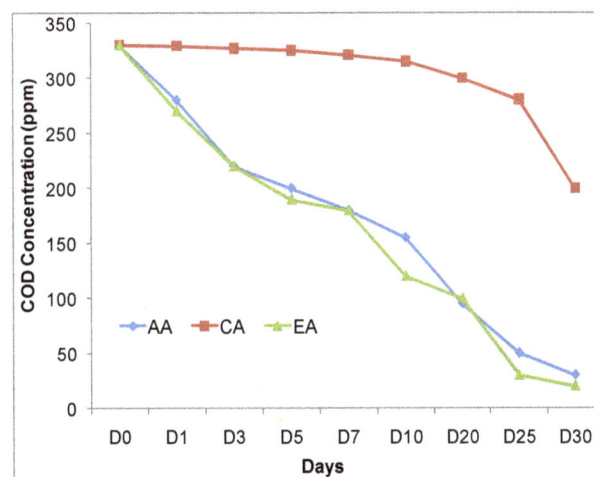

Figure 16. Variation of COD (ppm) in anaerobic pond

Figures 7, 8, 9 and 10, show reduction potential of NO_3 and NH_4 in Facultative and Anaerobic pond. Reduction in NO_3 and NH_4 concentration is a little bit slow. Significant reduction is observed in 15 days rather than 7 days as in case of NO_2 reduction. Figures 11 and 12, show macrophytes' potential of PO_4 absorption. In facultative pond, all the macrophytes show high potential of PO_4 absorption. Nearly 80% is absorbed in first 4 days only. However, in anaerobic pond simple containers, it took much longer time to absorb PO_4, 15 days or more. Absorption pattern is almost linear with time. Figures 13 and 14, shows variation of pH concentration in facultative and anaerobic pond samples. Variation in pH is not significant and shows a little erratic behavior. *Pistia stratiotes* and *Nymphaea rubra reduces* pH from 9.75 to 7.5. However in anaerobic pond samples, there is hardly any change visible.

Figure 15 shows variation of salinity in anaerobic condition, which is increasing. However, BOD concentration changes significantly, from 335 to less than 40 ppm (Figure 16). However it needed a period of 25 days. It is quite possible that in 25 days, BOD might have settled down as evident from indicator samples. However, still, macrophytes contribution of reducing BOD cannot be ignored.

4. Conclusion

Macrophytes (*Ipomoea aquatic, Trapa, Nymphaearubra,* and *Pistia* sp.) show robustness in removal of nutrient –pollutants for gainful application. Their effectiveness is depends types of pollutants, which provides leverage for waste management. *Ipomoea aquatic* and *Pistia* sp. are more effective in reducing pollutants from wastewater of municipal area.

References

Brix, H. (1994). Use of constructed wetlands in water pollution control: Historical development, present status, and future perspectives. *Water Science and Technology, 30*(8), 209-223.

Busnardo, M. J., Gersberg, R. M., Langis, R., Sinicrope, T. L., & Zedler, J. B. (1992). Nitrogen and phosphorus removal by wetland mesocosms subjected to different hydroperiods. *Ecological Engineering, 1*, 287-307. http://dx.doi.org/10.1016/0925-8574(92)90012-Q

Debusk, T. A., Reddy, K. R., & Clough K. S. (1989). Effectiveness of mechanical aeration in floating aquatic macrophytes-based wastewater treatment systems. *Journal of Environmental Quality, 18*(3), 349-354. http://dx.doi.org/10.2134/jeq1989.00472425001800030019x

Etnier, C., & Guterstam, B. (1996). *Ecological engineering for waste water treatment.* New York: CRC Press.

Gumbricht, T. (1993). Nutrient removal process in freshwater submersed macrophytes system. *Ecological Engineering, 2*(1), 1-30. http://dx.doi.org/10.1016/0925-8574(93)90024-A

Kadlec, R. H., & Knight, R. L. (1996). *Global Wetlands.* Florida : CRC Press.

Knight, R. L., Kadlec, R. H., & Reed, S. C. (1992). Wetlands for wastewater treatment database. *Proceedings of 3rd International Conf. on Wetland Systems for Water Pollution Control.* Sydney, Australia.

Mitsch, W. J. (1996). *Ecological engineering: a new paradigm for engineers and ecologists: Engineering within ecological constraints.* Washington DC: National Academy Press.

Mohanty, R. K., & Sinha, M. K. (1999). Use of aquatic macrophytes in water quality management. *Fishing chimes, 18*, 33-34.

Ozimek, T., Van Donk, E., & Gulati, R. D. (1993). Growth and nutrient uptake by two species of Elodea in experimental conditions and their role in nutrient accumulation in a macrophytes dominated lake. *Hydrobiologia, 251*, 13-18. http://dx.doi.org/10.1007/BF00007159

Poole, W. (1996). *Natural wastewater treatment with duckweed aquaculture, recycling resources, Ecological engineering for waste water treatment.* Environment Research Forum, vols. 5-6, Switzerland: Transtec Publications.

Welch, E. B. (1996). *Ecological Effects of Wastewater: Applied limnology and pollutant effects.* London: Chapman and Hall.

Yan. J., & Ma, S. (1991). The function of ecological engineering in environmental conservation in environmental conservation with some case studies from china. *Ecological engineering for waste water treatment.* Sweden: Gothenberg.

16

Evaluation of Heavy Metal in Soils From Enyimba Dumpsite in Aba, Southeastern Nigeria Using Contamination Factor and Geo-Accumulation Index

Amadi Akobundu N.[1] & Nwankwoala H. O.[2]

[1] Department of Geology, Federal University of Technology, Minna, Nigeria

[2] Department of Geology, University of Port Harcourt, Port Harcourt, Nigeria

Correspondence: Nwankwoala H. O., Department of Geology, University of Port Harcourt, Port Harcourt, Nigeria. E-mail: nwankwoala_ho@yahoo.com

Abstract

The manner in which municipal wastes generated are disposed in most urban areas in Nigeria is worrisome. The upsurge in population density and its resultant increase in urbanization and industrialization and the amount of waste generated in Aba, are of great concern. The objective of this research is to evaluate the concentration of some heavy metals in soils in the vicinity of Enyimba dumpsite in Aba, Nigeria. Thirty soil samples were collected and analyzed in the laboratory for some heavy metals by atomic absorption spectrophotometric method and multivariate statistical techniques. Twenty-five of the samples were obtained from the vicinity of the dumpsite while five samples are collected far away from the dumpsite to serve as control samples. The overall decreasing metal concentration in the dumpsite soil is: Cd > Co > Cu > Zn > As > Pb > Mn > Ni > Cr. A positive correlation exists between Cd and organic matter (r = 0.598). Geo-accumulation index and contamination factor showed a moderate contaminated with Cd only while the other metals are in their uncontaminated level. Factor analysis revealed four major components accounting for 78.82% of cumulative variance of the contamination: Cd, Cu, Co and organic matter; Pb, Zn and pH; Mn, As, clay + silt and finally Cr and Ni. From the above observations, it is evident that only Cd showed more pronounced level of pollution than any other metal. The need to replace open dumpsites with well designed sanitary landfills is advocated.

Keywords: heavy metals, contamination, dumpsite, analysis and Aba

1. Introduction

Open dumps are the oldest and most common way of disposing of solid wastes. The practice of landfill as a method of waste disposal in many developing countries is far from standard recommendations (Mull, 2005; Adewole, 2009). Solid and fluid wastes generation and their poor disposal mechanism in the urban areas of most developing countries have become a threat to the environment (Amadi et al., 2010). Rapid rural-urban migration and upsurge in population of many African, Asian and South American countries have also intensified and contributed their quota to the pollution hazards on and in the environment (Awomeso et al., 2010). Inadequate information and technology as well as insufficient resources and poor policy execution capacity are some of the causes of environmental pollution arising from municipal waste in most state capital in Nigeria. According to Amadi et al. (2010), dumpsites in most developing countries are usually unlined shallow hollow excavations arising from abandoned burrow-pits and quarry-sites without any environmental impact assessment studies.

Many cities in Nigeria have developed without proper planning and it has led to the presence of open dumps within built-up areas inhabited by millions of people. Consequently, such waste dumps become point source for soil pollution as they serve as host for leachate from dumpsites. The composition of solid wastes in major cities in Nigeria comprises domestic garbage, wood, agricultural waste, industrial waste, hospital waste, polythene bags, plastics, broken glasses, abandoned automobiles, demolition waste, ash, dust, human and animal waste. Solid waste are materials discarded after it has served its purpose or is no longer useful while industrial solid waste are usually by-product or end-product of materials from large-scale production factories and industries (Awomeso et al., 2010).

Due to the high cost of fertilizer, it is now a common practice for farmers to search for soils rich in organic manure. Such soils are easily obtained from dumpsites and used for planting of vegetables and food crops. Ademoroti (1990) ascertained that there is a positive linear correlation between heavy metals (Cd, Pb, and Ni) in the soil and vegetables grown on it. It has also been established that heavy metals have a high affinity for organic matter and clay soils (Bodur & Ergin, 1994; Zonta et al., 1994). The aim of this study was to examine the extent of heavy metal contamination in soils within Enyimba dumpsite. It also attempts to ascertain the suitability of such soils for agricultural purposes.

2. Materials and Methods

2.1 Study Area Description

Enyimba dumpsite is located in Aba, the commercial and industrial nerve centre of Southern Nigeria. The high number of markets, industries and fabricating companies in the area has resulted to high population density and high accumulation of wastes (Figure 1). Enyimba dumpsite lies between latitudes $05°06.796'$ N and $05°06.948'$ N and Longitudes $07°19.604'$ E and $07°19.758'$ E. The burrow-pits excavated during the construction of the Aba-Port-Harcourt expressway gave rise to Enyimba dumpsite. It is an open dumpsite and its proximity to markets and industries in Aba gives it high patronage. The heavy anthropogenic activities and the corresponding huge amount of wastes generated and discarded (Figure 1) on daily basis lead to the choice of Enyimba dumpsite for this study. Scavengers, birds, rodents, reptiles and micro-organisms abound in the decaying portions of the dump. The area is a low land and is drained by Imo and Aba Rivers and their tributaries (Figure 2). The area has two distinct seasons: a dry season which lasts from November to March, and a rainy season which starts from April to October. Rainfall is brought by the moist Equatorial Maritime Air Mass from the Gulf of Guinea with prevailing winds from the south to west. The average annual rainfall is about 2500 mm (Uma, 1990).

2.2 Geology and Hydrogeology

The study area is underlain by the Benin Formation which consists of unconsolidated, dominantly sandy formations also known as the coastal plain-sand of Miocene to Recent age (Uma, 1990). The formation is made up of very friable sands while clays occurring as streak and discontinuous lenses (Figure 2). Generally the sands are fine grained to coarse grained and are poorly sorted with pebble beds occurring in lenses (Onyeagocha, 1980). The studied area is underlain by a thick unconfined aquifer of regional extent. Lenticular clays and shales confine high yielding aquifers. Most of the boreholes tap unconfined aquifers which are regional in extent but comes in contact with Ogwashi-Asaba Formation and Alluvium in the north and south respectively (Figure 2).

Figure 1. An overview of Enyimba dumpsite (Source: Amadi, 2011)

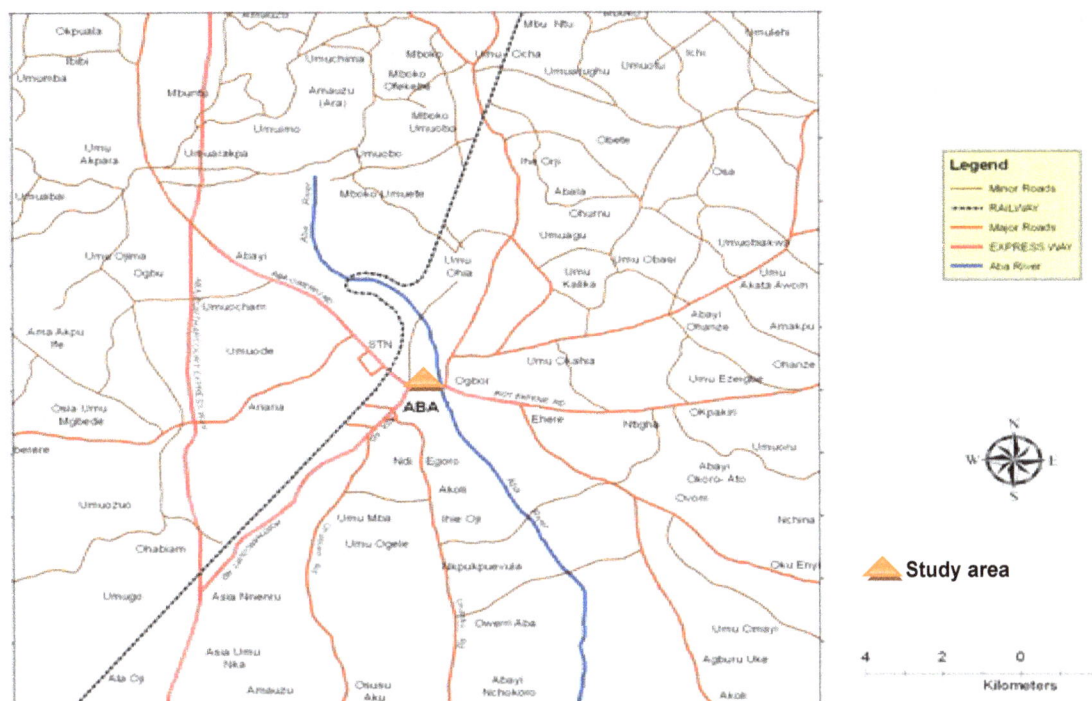

Figure 2. Location map of the study area (Source: Amadi, 2011)

2.3 Soil Sampling

A total of thirty soil samples were used for the present study. Twenty-five samples were collected within the vicinity of the dumpsite while five samples were collected far away from the dumpsite, which serves as control samples. Six samples were collected consecutively every month for five months, during the dry season, from November, 2007 to March, 2008. Samples collected were stored in sealed polythene bags and transported to the laboratory for pre-treatment and analyses.

2.4 Laboratory Analyses

The soil samples were air-dried, mechanically grounded using a stainless steel roller and sieved to obtain < 2 mm fraction. A 30 g sub-sample was taken from the original bulk soil of < 2 mm fraction and regrounded to obtain < 200 µm fraction using a mortar and pestle. This fine material was used to determine organic carbon and total metal content in soil. The < 2 mm fraction was used to determine pH (1: 5) soil/water extract and particle size analysis using Rayment and Higginson (1992) method. Organic carbon was determined by the modified Walkley and Black method by Saharawat (1982).

Soil samples were digested in a mixture of concentrated nitric acid (HNO_3), concentrated hydrochloric acid (HCl) and 27.5% hydrogen peroxide (H_2O_2) according to the USEPA method 3050B for the analysis of heavy metals (USEPA, 1996). A reagent blank was run for the set of six samples. The extracts were analyzed by atomic absorption spectrophotometer (Perkin Elmer, Model No. 2380).

2.5 Statistical Analysis

In order to quantitatively analyze and confirm the relationship among soil properties (pH, organic matter, clay and silt) and heavy metal content, Pearson correlation analysis was applied to the dataset. Principal component analysis (PCA) was adopted to assist the interpretation of elemental data. PCA was used in identifying the different groups of metals that correlate and thus can be considered as having a similar behavior and common source (Tahri et al., 2005). A component with an eigenvalues of less than one is considered less important and such an observed variable can be ignored (Baptista et al., 2007). All the statistical analyses were performed using SPSS for windows (release ver.11, Inc., Chicago, IL).

2.6 Data Analysis

Contamination factor (CF) and geo-accumulation index (GeoI) are quantitative check used to describe concentration trend of metals in soils. Contamination factor (CF) is a quantifier of the degree of contamination relative to either the average crustal composition of the respective metal or to measured background values from geologically similar and uncontaminated area (Tijani et al., 2004). It is expressed as:

$$CF = C_m / B_m$$

Where C_m is the mean concentration of metal m in soil and B_m is the background concentration (value) of metal m, either taken from the literature (average crustal abundance) or directly determined from a geologically similar material.

Geo-accumulation index (GeoI) as proposed by Mueller (1979) and cited by Lokeshwari and Chandrappa (2006) have been widely used to evaluate the degree of heavy metal contamination in terrestrial and aquatic environments as expressed:

$$GeoI = ln \ [C_m / 1.5 * B_m)$$

Where C_m and B_m are as defined above, while 1.5 is a factor for possible variation in the background concentration due to lithologic differences. GeoI is classified into seven descriptive classes as follows: < 0 = practically uncontaminated; 0-1 uncontaminated to slightly contaminated, 2-3 = moderately to highly contaminated, 4-5 = highly to very strongly contaminated, > 5 = very strongly contaminated. The latter is an open-end class that is indicative of all values greater than 5, and a GeoI of 6 is said to be indicative of 100-fold enrichment of a metal with respect to the baseline value (Mueller, 1979).

3. Results

The statistical summary of the analyzed heavy metals are contained in Table 1 while their respective mean concentration are displayed in Figure 3. The determined concentrations of contamination factor and geo-accumulation index of metals in dumpsite soil are shown in Table 2. Also, the results of correlation analysis and principal component analysis are summarized in Tables 3 and 4 respectively.

Table 1. Statistical summary of Heavy Metals concentration in (mg/kg)

Parameters	Minimum	Maximum	Mean
Cd	0.18	2.60	1.40
Mn	0.30	48.10	28.50
Cu	1.06	15.98	12.86
Cr	0.02	2.78	1.34
Ni	0.50	6.25	2.94
Pb	0.24	2.15	1.08
As	0.01	0.08	0.05
Zn	2.40	28.50	16.04
Co	0.20	17.90	10.58
pH	4.80	6.90	5.70

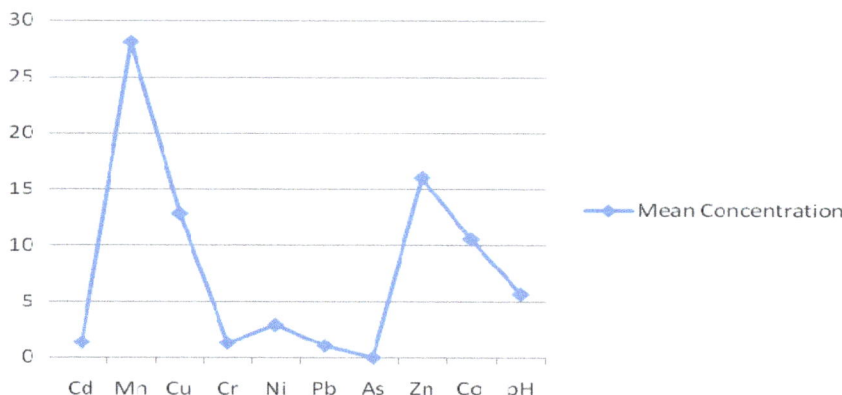

Figure 3. Mean concentration of heavy metals from Eyimba dumpsite, Aba

Table 2. Metal contamination factor and geo-accumulation index of metals in soil from the dumpsite

Parameters	C_m	B_m	CF	GeoI	Overall summary of contamination level
Cd	1.40	0.15	9.33	1.828	Moderately contaminated
Mn	48.12	1000	0.048	-3.442	Uncontaminated
Cu	12.86	70	0.184	1.098	Slightly contaminated
Cr	1.34	122	0.011	-4.920	Uncontaminated
Ni	2.94	80	0.037	-3.709	Uncontaminated
Pb	1.08	16	0.068	-3.101	Uncontaminated
As	0.05	5	0.010	-3.007	Uncontaminated
Zn	16.04	132	0.122	-2.513	Uncontaminated
Co	10.58	23	0.460	-1.181	Uncontaminated

CF- contamination factor; GeoI- geo-accumulation index;

C_m- mean concentration of the metal in the soil;

B_m- average crustal abundance (background value) in an uncontaminated soil, adopted from (Dineley et al., 1976).

Table 3. Pearson correlation coefficient matrix for heavy metals in soils from the dumpsite

	Cd	Mn	Cu	Cr	Ni	Pb	Ar	Zn	Co	pH	OM	C + S
Cd	1.000											
Mn	0.109	1.000										
Cu	0.065	-0.112	1.000									
Cr	0.252	0.041	0.141	1.000								
Ni	0.354	0.678**	0.101	0.093	1.000							
Pb	0.327	-0.113	0.818**	0.008	0.334	1.000						
As	0.080	0.199	0.249	0.118	-0.333	0.090	1.000					
Zn	0.153	0.205	0.788**	-0.044	0.534*	0.637**	0.110	1.000				
Co	0.433*	0.084	0.211	0.208	0.360*	0.016	0.186	0.127	1.000			
pH	0.106	-0.112	0.024	0.091	0.119	0.095	0.112	0.085	0.101	1.000		
OM	0.598*	0.724**	0.028	0.284	0.284	0.195	0.220	0.054	0.066	0.841*	1.000	
C + S	0.045	0.293	0.123	0.031	-0.023	0.545*	0.151	0.049	0.137	0.521	-0.192	1.000

**: Correlation is significant at the 0.01 level (2-tailed); *: Correlation is significant at the 0.05 level (2-tailed); OM: Organic Matter; C + S: Clay + Silt.

Table 4. Varimax normalized rotated principal component loading of selected metals and soil components

Variables	PC-1	PC-2	PC-3	PC-4
Cd	**0.809**	0.320	0.056	-0.207
Mn	-0.072	-0.027	**0.858**	0.204
Cu	**0.605**	0.341	-0.074	0.082
Cr	-0.340	0.178	0.151	**0.502**
Ni	0.123	0.109	-0.148	**0.613**
Pb	0.310	**0.633**	-0.117	-0.020
As	-0.134	0.029	**0.580**	0.231
Zn	0.228	**0.734**	0.163	0.530
Co	**0.735**	-0.423	-0.193	0.745
pH	0.159	**0.631**	0.072	0.231
Organic Matter	**0.720**	-0.193	-0.088	-0.195
Clay + Silt	-0.203	-0.145	**0.865**	0.136
Eigenvalue	4.142	3.078	2.641	1.705
Total Variance (%)	26.356	20.321	17.785	14.354
Cumulative %	26.356	46.677	64.462	78.816

4. Discussion

The fieldwork was done during the dry season in order to obtain maximal heavy metal concentration from the soil. Yahaya (2009) confirmed that the concentration of heavy metal in soil is higher in dry season than in rainy season because more heavy metals are lost in the soil due to run-off and infiltration in rainy season which are absent in dry season. The concentration of cadmium ranges from 0.18-2.60 mg/kg with a mean concentration of 1.40 mg/kg (Table 1). The values of Cd obtained in this study are higher than the average crustal abundance of 0.15 ppm in an uncontaminated soil. The calculated geo-accumulation index (GeoI) for cadmium indicates that the soils around the dumpsite are moderately contaminated (Table 2) and Cd showed moderately positive correlation with Cobalt and organic matter (0.05 level).

Cadmium metal is used as an anticorrosive, electroplated on steel, Cadmium sulfide and selenide are commonly used as pigments in plastics, batteries and in various electronic components. It is also used with inorganic fertilizers produced from phosphate ores and when these products are no more servisable, they are thrown into the dump as waste. During decomposition, the Cd component is leached into the surrounding soil and over time gets accumulated in the soil. Cadmium is extremely toxic and the primary use of soil high in Cd in form of manure for the cultivation of vegetables and other food crops could cause adverse health effect to consumers such as renal disease and cancer (Che et al., 2003; Gorenc et al., 2004). Moreover, when ingested by humans, cadmium accumulates in the intestine, liver and kidney and chronic exposure of Cd causes proximal tubular disease and osteomalacia (Pascual et al., 2004). Therefore, the soils from this dumpsite are not suitable for agricultural purposes.

Manganese ranged 0.30-92.10 mg/kg. The mean was 48.12 mg/kg. Abbasi et al. (1998) gave an accepted value of 1000 mg/kg for manganese in an uncontaminated soil and the calculated GeoI value gave a value that indicates uncontaminated. Manganese is essential for plants and animals. Manganese dioxide and other manganese compounds are used in products such as batteries, glass and fireworks (Huang & Lin, 2003; Aboud & Nandini, 2009). Potassium permanganate is used as an oxidant for cleaning, bleaching and disinfection purposes. Other manganese compounds are used in fertilizer, fungicides and as livestock feeding supplements. It can be adsorbed onto soil depending on organic content, pH, grain-size and cation exchange capacity (CEC) of the soil and this can be exemplified by the strong positive correlation (Table 3) with organic matter (< 0.01 level).

Concentration in copper varied from 1.06-15.98 mg/kg with an average value of 12.86 mg/kg. A moderately high positive correlation with lead and Zinc was established (< 0.01 level). Copper is widely used in electrical wiring, roofing, various alloys, pigments, cooking utensils, piping and in the chemical industries (Aboud & Nandini, 2009). Copper compounds are used in fungicides, algicides, insecticides, wood preservation, electroplating, dye manufacture, engraving, lithography, petroleum refining and pyrotechnics. It is also added to fertilizers and animal feeds as a nutrient to support plant and animal growth (Mielke et al., 1991; Pascual et al., 2004). The Cu concentration in GeoI is within the uncontaminated level.

Chromium concentration ranges from 0.02-2.78 mg/kg with a mean value of 1.34 mg/kg. No correlation was found with other metals and its concentration falls within the uncontaminated. It is used in alloys, electroplating, pigments, paints manufacture, fungicides, photography, glass and leather tanning industries. Chromium is carcinogenic by inhalation and corrosive to tissue (Lin et al., 2002; Aboud & Nandini, 2009).

Nickel measured concentrations are below the average crustal abundance in an uncontaminated soil. A moderate positive correlation with Zn was noted at < 0.05 level (Table 3). Nickel is used mainly as alloys, which are characterized by their hardness, strength, and resistance to corrosion and heat. It is a major component in the production of stainless steels, non-ferrous alloys and super alloys. Other application of Ni includes electroplating, as catalysts, in nickel-cadmium batteries, coins, welding and electronic products (Pascual et al., 2004; Amadi, 2011).

The results show that lead concentration deposited at the dumpsite ranged 0.24-2.15 mg/kg with a mean concentration of 1.08 mg/kg (Table 1). Though there was an observed strong correlation with Cu (< 0.01 level), its concentration is within the level of uncontaminated soil. Lead is non essential for plants and animals and is toxic by ingestion-being a cumulative poison (MacFarlane & Burchett, 2002; Sharma & Pervez, 2003). Lead toxicity leads to anaemia both by impairment of haemo-biosynthesis and acceleration of red blood cell destruction. In addition, Pb reduces sperm count, damages kidney, liver, blood vessels, nervous system and other tissues in human (Anglin-Brown et al., 1995). Other uses of lead is in the production of lead acid batteries, solder, alloys, cable sheathing, pigments, ammunition, glass and plastic stabilizers. Tetraethyl and tetramethyl lead are important due to their extensive use as antiknock compounds in petrol (Mielke et al., 1991; McAllister et al., 2005).

Arsenic concentration varied between 0.01 mg/kg and 0.08 mg/kg with an average concentration of 0.05 mg/kg. These values are found to be low the critical value of 16 mg/kg (average crustal abundance) for an uncontaminated soil (Table 2). The GeoI concentration lies below the range for uncontaminated soil. Arsenic is highly carcinogenic has no nutritional value for plant and animal (Amadi et al., 2010).

Zinc in the study ranged 2.40-28.50 mg/kg. The mean value was 16.04 mg/kg. With this values, the concentration of Zn in soils from the dumpsite are within the stipulated guideline limits (Table 2). Zinc had very strong positive correlation with Cu and Pb (< 0.01 level) and moderately positive correlation with Ni (< 0.05 level). It is an essential growth element for plants and animals but can be toxic at elevated concentration. Zinc is used in making alloys of brass and bronze, batteries, fungicides, pigments, pesticides, galvanizing steel and iron products. It is used in combination with some enzymes system which contributes to energy metabolism, transcription and translation (Anglin-Brown et al., 1995). Excessive concentration of Zn in soil leads to phyto-toxicity as it is a weed killer (Preda & Cox, 2002; Aboud & Nandini, 2009).

Cobalt concentration ranged 0.20-17.90 mg/kg with a mean value of 10.57 mg/kg. The measured concentrations of Co are acceptable range for an uncontaminated soil (Table 2). Cobalt is widely used as alloys for steels, electroplating, fertilizer, porcelain and glass making. It is essential for the growth of algae and bacteria but required in trace concentration for higher plants and animals (Mielke et al., 1991; Rayment & Higginson, 1992; Aboud & Nandini, 2009; Amadi et al., 2012).

Among significant variables that control the distribution and enrichment of heavy metals in soils are pH of soil, grain size of the soil, amount of organic matter in the soil and the cation exchange capacity of the soil (Lin et al., 2002; Huang & Lin, 2003). The soil pH is generally low, signifying acidic soil while loamy soil characterize the top soil at the dumpsite and these condition enhances the precipitation and bio-accumulation of heavy metals in soil (Ujevic et al., 2000). Heavy metals have a strong affinity for organic content, clay and silt fraction because of their high cation exchange capacity (Bodur & Ergin, 1994; Zonta et al., 1994). The top-soil from the dumpsite comprises of organic content, clay and silt fraction. This agrees with the result of geophysical investigation carried out earlier which suggest the presence leachate near the top-soil (Amadi et al., 2010)

Four principal components (Eigenvalues > 1) emerged accounting for 78.82% of cumulative variance from the principal component analysis (Table 4). The first principal component (PC-1) loading with 26.36% variance showed higher loading for Cd, Cu, Co and organic matter. Human activities in the area involving electrical wiring, various alloys, alloys, pigments, fungicides, insecticides, electroplating, cooking utensils, batteries and dye production are the possible sources of Cd, Cu and Co. When these products are thrown into the dumpsite, these elements are leached away and accumulate at the top soil where they are adsorbed because of affinity for metals by organic matter (Rayment & Higginson, 1992; Odero et al., 2000; Amadi, 2011).

The second principal component (PC-2) has loading 20.32% of total variance, had high loading for Pb, Zn and pH. These might be due to soldering, battery charging, zinc-roofing sheet, electroplating cable sheathing, pigments, ammunition, glass and plastic stabilizers, artisanal activities going on in this area. Pb and Zn are essential components of the raw material used in soldering wires and lead accumulators (Odero et al., 2000; Banar et al., 2006). The pH of the soil could have contributed to Pb and Zn retention in the soil, resulting in low mobility of the metals (Alloway, 1990; Yoshida et al., 2002; Amadi et al., 2012).

The third principal component (PC-3) explains 17.79% of the total variance and comprises of Mn, As and clay plus silt. Industrial activities domiciled in the area may be responsible for the presence of Mn and As while the physico-chemical properties of clay could have encouraged their availability in the soil. The fourth principal component (PC-4) has a moderate loading for Cr and Ni which accounts for 14.35% of the total variance. This could be attributed to domestic waste discharged at the dumpsite and the decomposition of vehicle and machine scraps (Pereira et al., 2007). The dumping of unwanted portion of paints, fungicides, photographic films, glass and waste from leather tanning industries can also enrich the soil with Cr and Ni. The concentrations of heavy metals in the control samples are negligible, typical of an uncontaminated soil and this further confirmed that the dumpsite is a possible source of heavy metals in the soil.

5. Conclusion

In this study, contamination factor, geo-accumulation index, correlation and principal component analysis were used for determining the environmental quality of soils from dumpsite in terms of heavy metal accumulation and other soil properties. The result revealed the following trend in their order of geo-accumulation in the soil: Cd > Co > Cu > Zn > As > Pb > Mn > Ni > Cr. There is a very strong correlation between organic matter content on cadmium and copper metal accumulation, suggesting that the soil around the dumpsite are moderately and slightly contaminated with cadmium and copper respectively. Contamination factor and geo-accumulation index further confirmed that the soil from the dumpsite was moderately contaminated with Cd, slightly contaminated with Cu and presently uncontaminated with Co, Zn, As, Pb, Mn, Ni and Cr. The principal component analysis summarizes (reduces) the dataset into four major components representing four possible different sources of the elements. The effectiveness of multivariate statistical analysis in evaluating heavy metal concentration in dumpsite soils has been demonstrated in this study.

Recommendation

A well designed sanitary landfill that incorporates the local geology, prevalent climatic condition, slope geometry, type of waste generated, nature of settlement and cultural believe of the people that will mitigate (impede) the infiltration of the leachate into the soil and shallow groundwater system are advocated. The use of Enyimba dumpsite should be discontinued. Although no severe pollution may have occurred at present apart from cadmium, the continuous dumping of waste at the dumpsite may lead to the enrichment of the soil with other metals that are presently at uncontaminated levels. Therefore, separation and recycling of wastes as well as the use of sanitary landfills and incinerators should be encouraged. Due to the toxicity of heavy metals, the use of manure from the dumpsite for agricultural purposes should be discouraged as plants and vegetables can easily absorb them.

Acknowledgement

The researchers acknowledge their mentor, Prof P. I. Olasehinde, Department of Geology Department, Federal University of Technology, Minna, Nigeria for his support and encouragement.

References

Abbasi, S. A., Abbasi, N., & Soni, R. (1998). *Heavy metals in the environment* (1st ed., p. 314). Mittal Publ.

Aboud, S. J., & Nandini, N. (2009). Heavy metal analysis and sediment quality values in urban lakes. *Am. J. Environ. Sci., 5*(6), 678-687.

Ademoroti, C. M. A. (1990). Bio-accumulation of heavy metals in some Mangrove Fauna and Flora. In *Environmental chemistry and toxicological consultancy, Benni,* 180-182.

Adewole, A. T. (2009). Waste management towards sustainable development in Nigeria: A case study of Lagos State. *Int. NGO J., 4*(4), 173-179.

Alloway, B. J. (1990). *Heavy metals in soil* (p. 339). New York: John Wiley and sons Inc.

Amadi, A. N. (2010). Effects of urbanization on groundwater quality: A case study of Port-Harcourt, Southern Nigeria. *Natur. Appl. Sci. J., 11*(2), 143-152.

Amadi, A. N. (2011). Assessing the Effects of Aladimma Dumpsite on Soil and Groundwater Using Water Quality Index and Factor Analysis. *Australian Journal of Basic and Applied Sciences, 5*(11), 763-770.

Amadi, A. N., Ameh, M. I., & Jisa, J. (2010). The impact of dumpsites on groundwater quality in Markurdi Metropolis, Benue State. Natur. *Appl. Sci. J., 11*(1), 90-102.

Amadi, A. N., Olasehinde, P. I., Okosun, E. A., Okoye, N. O., Okunlola, I. A., Alkali, Y. B., & Dan-Hassan, M. A. (2012). A Comparative Study on the Impact of Avu and Ihie Dumpsites on Soil Quality in Southeastern Nigeria. *American Journal of Chemistry, 2*(1), 17-23.

Anglin-Brown, B., Armour, A., & Lalor, G. C. (1995). Heavy metal pollution in Jamaica 1: Survey of cadmium, lead and zinc concentrations in the Kintyre and Hope flat district. *Environ. Geochem. Health, 17*, 51-56.

Awomeso, J. A., Taiwo, A. M., Gbadebo, A. M., & Arimoro, A. O. (2010). Waste disposal and pollution management in urban areas: A workabale remedy for the environment in developing countries. *Am. J. Environ. Sci., 6*(1), 26-32.

Banar, M., Aysun, O., & Mine, K. (2006). Characterization of the leachate in an urban landfill by physicochemical analysis and solid phase microextraction. *GC/MS. Environ. Monitor. Assess., 121,* 439-459.

Baptista, N. J. A., Smith, B. J., & McAllister, J. J. (2007). Concentration of heavy metals in sediments from urban runoff: implications for environmental quality emNitero'I / RJ-Brazil. *An acad. Bras Cienc., 79,* 981-995.

Bodur, M. N., & Ergin, M. (1994). Geochemical characteristics of the recent sediment from the Sea of Marmara. *Chem. Geol., 115,* 73-101. http://dx.doi.org/10.1016/0009-2541(94)90146-5

Che, Y. Q., & Lin, W. Q. (2003). The distributions of particulate heavy metals and its indication to the transfer of sediments in the Changjiang estuary and Hangzhou Bay. *Mar. pollut. Bull., 46,* 123-131. http://dx.doi.org/10.1016/S0025-326X(02)00355-7

Dineley, D., Hawkes, D., Hancock, P., & Williams, B. (1976). *Earth resources – a dictionary of terms and concepts* (p. 205). London: Arrow Books Ltd.

Gorenc, S., Kostaschuk, R., & Chen, Z. (2004). Spatial variation in heavy metals on tidal flats in the Yangtze Estuary China. *Environ. Geo., 45,* 1101-1108.

Huang, K., & Lin, S. (2003). Consequences and implication of heavy metal spatial in sediments of Keelung River drainage basin, Taiwan. *Chemosp., 53,* 1113-1121.

Lin, Y. P., Teng, T. P., & Chang, T. K. (2002). Multivariate analysis of soil heavy metal pollution and landscape in Changhua Country in Taiwan. *Landscape Urban Plan., 62,* 19-35. http://dx.doi.org/10.1016/S0169-2046(02000094-4

Lokeshwari, H., & Chandrappa, G. T. (2006). Impact of heavy metals content in water, water hyacinth and sediments of Laibagh tank, Bangalore. *Indian J. Environ. Sci. Eng., 48,* 183-188. http://www.neeri.res.in/jesevo14803006.pdf

MacFarlane, G. R., & Burchett, M. D. (2002). Toxicity, growth and accumulation relationships of copper, lead and zinc in the Gray Mangrove Avicennia marina (Forsk) Veirh. *Marine Environ. Res., 54,* 65-85. http://dx.doi.org/10.1016/S0141-1136(02)00095-8.

McAllister, J. J., Smith, B. J., Baptista, N. J. A., & Simpson, J. K. (2005). Geochemical distribution and bioavailability of heavy metals and oxalate in street sediments from Rio de Janeiro, Brazil: A preliminary investigation. *Environ. Geoch. Heal., 27,* 429-441.

Mielke, H. W. (1994). Lead in New Orleans soils: new images of an urban environment. *Environ. Geochem. Health, 16,* 123-128.

Mueller, G. (1979). Schwermettale in den sedimenten des Rheins – Veraenderungen seit. *Umschau, 79,* 778-783.

Mull, E. J. (2005). Approaches toward sustainable urban solid waste management: Sahakaranagar Layout. Unpublished M.Sc. thesis, Int. Environ. Sci., Lund University, Lund, Sweden, p. 37.

Odero, D. R., Semu, E., & Kamau, G. (2000). Assessment of cottage industries-derived heavy metal pollution of soil within Ngara and Gikomba area of Nairobi city, Kenya. *Afri. J. Sci. Technol., 1,* 52-62.

Onyeagocha, A. C. (1980). Petrography and Depositional Environment of the Benin Formation. *Nig. J. Min. Geol., 17*, 147-151.

Pascual, B., Gold-Bouchot, G., Ceja-Moreno, V., & del Ri'o-garci'a, M. (2004). Heavy metal and hydrocarbons in sediments from three lakes from san Miguel, Chiapas, Mexico. *Bull. Environ. Contam. Toxicol., 73*, 762-769.

Perira, E., Baptista-Nato, J. A., Smith, B. J., & Mcallister, J. J. (2007). The contribution of heavy metal pollution derived from highway runoff to Guanabara Bay sediments-Rio de Janeiro? *Brazil Annals Brazillian Acad. Sci., 79*, 739-750.

Preda, M., & Cox, M. E. (2002). Trace metal occurrence and distribution in sediments and mangroves, Pumicestone region, southeast Queenland, Australia. *Environ. Int., 28*, 433-449.

Rayment, G. E., & Higginson, F. R. (1992). *Australian Laboratory Handbook of soil and water chemical methods*. Port Melbourne: Reed International books Australia P/L, trading as Inkata Press.

Saharawat, K. L. (1982). Simple modification of the Walkey-Black method for simultaneous determination of organic carbon and potentially mineralizable nitrogen in tropical rice soils. *Plant and soils, 69*, 73-77.

Sharma, R., & Pervez, S. (2003). Enrichment and exposure of particulate lead in a traffic environment in India. *Environ. Geochem. Health, 25*, 297-306.

SPSS-16. (2009). *Statistical Package for the Social Sciences*. Chicago, USA: SPSS Inc.

Tahri, F., Benya, M., Bounakla, E. I., & Bilal, J. J. (2005). Multivariate analysis of heavy metal in soils, sediments and water in the region of Meknes, Central morocco. *Environ. Monitor. Asses., 102*, 405-417.

Tijani, M. N., Jinno, K., & Hiroshiro, Y. (2004). Environmental impact of heavy metal distribution in water and sediment of Ogunpa River, Ibadan area, southwestern Nigeria. *J. Min. Geol., 40*(1), 73-83.

Uma, K. O. (1989). Water resource of Owerri and its environs, Imo state, Nigeria. *J. Min. Geol., 22*(1-2), 57-64.

USEPA. (1996). Test methods for evaluating solid waste. *Physical/Chemical Methods* (3rd ed.). Method 3050B, Acid Digestion of Sediment, Sludges and soils, USEPA, Washington, DC, SW-846.

Yahaya, M. I., Mohammad, S., & Abdullahi, B. K. (2009). Seasonal variation of heavy metal concentration in Abattoir dumpsite soil in Nigeria. *J. Appl. Sci. and Envr. Mgt., 13*(4), 9-13.

Yoshida, M., Ahmed, S., Nebil, S., & Ahmed, G. (2002). Characterization of leachate from Henchir El Yahoidia close landfill. *Water Waste Environ. Res., 1*, 129-142.

Zonta, R., Zaggia, L., & Argese, E. (1994). Heavy metal and grain-size distribution in estuarine shallow water sediments of the Cona Marsh, Venice Lagoon, Italy. *Sci. Total Environ., 151*, 19-28. http://dx.doi.org/10.1016/0048-9697(94)90482-0

Permissions

List of Contributors

R. A. Dunlap
Department of Physics and Atmospheric Science, Dalhousie University, Halifax, Nova Scotia, Canada

Takuya Ito
Department of Materials and Life Science, Seikei University, Tokyo, Japan

Kazuyuki Yamada
Department of Materials and Life Science, Seikei University, Tokyo, Japan

Sigeru Kato
Department of Materials and Life Science, Seikei University, Tokyo, Japan

Hideki Suganuma
Department of Materials and Life Science, Seikei University, Tokyo, Japan

Akihiro Yamasaki
Department of Materials and Life Science, Seikei University, Tokyo, Japan

Seiichi Suzuki
Department of Materials and Life Science, Seikei University, Tokyo, Japan

Toshinori Kojima
Department of Materials and Life Science, Seikei University, Tokyo, Japan

Mochamad Syamsiro
Department of Environmental Science and Technology, Tokyo Institute of Technology, Yokohama, Japan

Wu Hu
Department of Environmental Science and Technology, Tokyo Institute of Technology, Yokohama, Japan

Shuta Komoto
Department of Environmental Science and Technology, Tokyo Institute of Technology, Yokohama, Japan

Shuo Cheng
Department of Environmental Science and Technology, Tokyo Institute of Technology, Yokohama, Japan

Putri Noviasri
Department of Environmental Science and Technology, Tokyo Institute of Technology, Yokohama, Japan

Pandji Prawisudha
Department of Mechanical Engineering, Institut Teknologi Bandung, Bandung, Indonesia

Kunio Yoshikawa
Department of Environmental Science and Technology, Tokyo Institute of Technology, Yokohama, Japan

Peter Andráš
Faculty of Natural Sciences, Matej Bel University, Banská Bystrica, Slovakia
Geological Institute of Slovak Academy of Sciences, Banská Bystrica, Slovakia

Ingrid Turisová
Faculty of Natural Sciences, Matej Bel University, Banská Bystrica, Slovakia

Eva Lacková
VŠB-Technical University of Ostrava, Ostrava, Czech Republic

Sherif Kharbish
Geology Department, Faculty of Science, Suez University, Suez Governate, El Salam City, Egypt

Jozef Krnáč
Faculty of Natural Sciences, Matej Bel University, Banská Bystrica, Slovakia

Lenka Čmielová
VŠB-Technical University of Ostrava, Ostrava, Czech Republic

Oluseyi E. Ewemoje
Department of Agricultural and Environmental Engineering, University of Ibadan, Ibadan, Nigeria

Samuel O. Ihuoma
Department of Agricultural and Environmental Engineering, University of Ibadan, Ibadan, Nigeria

Carlos M. Morales-Bautista
División Académica de Ciencias Biológicas (DACBiol.), Universidad Juárez Autónoma de Tabasco, Villahermosa, Tabasco, Mexico

Randy H. Adams
División Académica de Ciencias Biológicas (DACBiol.), Universidad Juárez Autónoma de Tabasco, Villahermosa, Tabasco, Mexico

Francisco Guzmán-Osorio
División Académica de Ciencias Biológicas (DACBiol.), Universidad Juárez Autónoma de Tabasco, Villahermosa, Tabasco, Mexico

Deysi Marín-García
División Académica de Ciencias Biológicas (DACBiol.), Universidad Juárez Autónoma de Tabasco, Villahermosa, Tabasco, Mexico

Almaz Akhmetov
ENCA Management, Yessik, Kazakhstan
Orizon Consulting Services, McLean, VA, USA

Ashraf S. Elkady
Department of Physics, Faculty of Science for Girls, King Abdulaziz University, Jeddah, KSA
Department Reactor Physics, NRC, Egyptian Atomic Energy Authority (EAEA), Cairo, Egypt

Walaa M. Abdel-Aziz
Department Reactor Physics, NRC, Egyptian Atomic Energy Authority (EAEA), Cairo, Egypt

Ibrahim I. Bashter
Department of Physics, Faculty of Science, Zagazig University, Zagazig, Egypt

Sunbong Lee
Department of Environmental Science and Technology, Interdisciplinary Graduate School of Science and Engineering, Tokyo Institute of Technology, Kanagawa, Japan

Koji Yoshida
Department of Mechanical Engineering, College of Science and Technology, Nihon University, Tokyo, Japan

Kunio Yoshikawa
Department of Environmental Science and Technology, Interdisciplinary Graduate School of Science and Engineering, Tokyo Institute of Technology, Kanagawa, Japan
Department of Mechanical Engineering, College of Science and Technology, Nihon University, Tokyo, Japan

Adrian K. James
Natural Resource and Environmental Science, University of Northern British Columbia, Prince George, BC, Canada

Steve S. Helle
Environmental Science & Engineering, University of Northern British Columbia, Prince George, BC, Canada

Ronald W. Thring
Environmental Science & Engineering, University of Northern British Columbia, Prince George, BC, Canada

P. Michael Rutherford
Environmental Science & Engineering, University of Northern British Columbia, Prince George, BC, Canada

Mohammad S. Masnadi
Chemical and Biological Engineering, University of British Columbia, Vancouver, BC, Canada

S. C. Nwanya
Department of Mechanical Engineering, University of Nigeria, Nsukka, Nigeria

I. Offili
Projects Development Institute (PRODA), Emene, Enugu, Nigeria

Randy L. Maddalena
Lawrence Berkeley National Laboratory, Berkeley, CA, USA

Melissa M. Lunden
Lawrence Berkeley National Laboratory, Berkeley, CA, USA

Daniel L. Wilson
Lawrence Berkeley National Laboratory, Berkeley, CA, USA
University of California, Berkeley, CA, USA

Cristina Ceballos
Lawrence Berkeley National Laboratory, Berkeley, CA, USA
University of California, Berkeley, CA, USA

Thomas W. Kirchstetter
Lawrence Berkeley National Laboratory, Berkeley, CA, USA

Jonathan L. Slack
Lawrence Berkeley National Laboratory, Berkeley, CA, USA

Larry L. Dale
Lawrence Berkeley National Laboratory, Berkeley, CA, USA

Yurii Maletin
YUNASKO-Ukraine, Kiev, Ukraine
Institute for Sorption and Problems of Endoecology, National Academy of Science of Ukraine, Kiev, Ukraine

Volodymyr Strelko
Institute for Sorption and Problems of Endoecology, National Academy of Science of Ukraine, Kiev, Ukraine

Natalia Stryzhakova
YUNASKO-Ukraine, Kiev, Ukraine
Institute for Sorption and Problems of Endoecology,
National Academy of Science of Ukraine, Kiev, Ukraine

Sergey Zelinsky
YUNASKO-Ukraine, Kiev, Ukraine
Institute for Sorption and Problems of Endoecology,
National Academy of Science of Ukraine, Kiev, Ukraine

Alexander B. Rozhenko
YUNASKO-Ukraine, Kiev, Ukraine

Denis Gromadsky
YUNASKO-Ukraine, Kiev, Ukraine

Vitaliy Volkov
Institute of Chemical Physics, Russian Academy of
Science, Chernogolovka, Moscow Region, Russia

Sergey Tychina
YUNASKO-Ukraine, Kiev, Ukraine
Institute for Sorption and Problems of Endoecology,
National Academy of Science of Ukraine, Kiev, Ukraine

Oleg Gozhenko
YUNASKO-Ukraine, Kiev, Ukraine
Institute for Sorption and Problems of Endoecology,
National Academy of Science of Ukraine, Kiev, Ukraine

Dmitry Drobny
YUNASKO-Ukraine, Kiev, Ukraine
Institute for Sorption and Problems of Endoecology,
National Academy of Science of Ukraine, Kiev, Ukraine

Neva Rebolj
University of Ljubljana, Faculty of Education, Kardeljeva
pl. 16, 1000 Ljubljana, Slovenia

Iztok Devetak
University of Ljubljana, Faculty of Education, Kardeljeva
pl. 16, 1000 Ljubljana, Slovenia

Sukhen Roy
Department of Environmental Management, University
of Kalyani, West Bengal, India

J. K. Biswas
Department of Environmental Management, University
of Kalyani, West Bengal, India

Sanjay Kumar
Department of Physics, B. R. Ambedkar Bihar University,
Muzaffarpur, Bihar, India
Centre for Renewable Energy and Environmental
Research, Muzaffarpur, Bihar, India

N. Amadi Akobundu
Department of Geology, Federal University of Technology,
Minna, Nigeria

H. O. Nwankwoala
Department of Geology, University of Port Harcourt, Port
Harcourt, Nigeria